Atmospheric Monitoring for High Energy AstroParticle Detectors (AtmoHEAD 2016)

EPJ Web of Conferences Volume 144 (2017)

Olomouc, Czech Republic
12 - 14 September 2016

Editors:

Petr Travnicek
Jan Ebr
Michael Prouza
Dusan Mandat

ISBN: 978-1-5108-4118-5

Printed from e-media with permission by:

Curran Associates, Inc.
57 Morehouse Lane
Red Hook, NY 12571

proceedings.com

Some format issues inherent in the e-media version may also appear in this print version.

Printed by Curran Associates, Inc. (2017)

For additional information, please contact EDP Sciences – Web of Conferences at the address below.

EDP Sciences – Web of Conferences
17, Avenue du Hoggar
Parc d'Activité de Courtabœuf
BP 112
F-91944 Les Ulis Cedex A
France

Phone: +33 (0) 1 69 18 75 75
Fax: +33 (0) 1 69 28 84 91

contact-edps@webofconferences.org

Additional copies of this publication are available from:

Curran Associates, Inc.
57 Morehouse Lane
Red Hook, NY 12571 USA
Phone: 845-758-0400
Fax: 845-758-2633
Email: curran@proceedings.com
Web: www.proceedings.com

TABLE OF CONTENTS

edp sciences EPJ Web of Conferences

Statement of Peer review

In submitting conference proceedings to EPJ Web of Conferences, I certify to the Publisher that I adhere to the **Policy on Publishing Integrity** of the journal in order to safeguard good scientific practice in publishing.

1. All articles have been subjected to peer review administered by the proceedings editors.
2. Reviews have been conducted by expert referees, who have been requested to provide unbiased and constructive comments aimed, whenever possible, at improving the work.
3. Proceedings editors have taken all reasonable steps to ensure the quality of the materials they publish and their decision to accept or reject a paper for publication has been based only on the merits of the work and the relevance to the journal.

Title, date and place of the conference

Atmospheric Monitoring for High Energy Astroparticle Detectors (ATMOHEAD) 2016,
12-14 September 2016, Olomouc, Czech Republic

Proceedings editor(s):

Petr Travnicek, Jan Ebr, Michael Prouza, Dusan Mandat

Date and editor's signature

18.4.2017

Preface

Dušan Mandát, *Jan* Ebr, *Michael* Prouza and *Petr* Trávníček

Institute of Physics of the Czech Academy of Sciences,
Na Slovance 2, Praha 8, Czech Republic

The AtmoHEAD (ATmospheric MOnitoring for High Energy Astroparticle Detectors) conference, Olomouc 12–14 September, 2016 was organized by the Palacky University Olomouc and the Institute of Physics of Academy of Sciences of the Czech Republic. The event follows the tradition of previous AtmoHEAD conferences – AtmoHEAD 2013 and 2014 in Saclay and Padova and ATMON 2008 and 2010 in Prague and Madison, and the first Astroparticles and Atmosphere (AA) conference held in 2003 Paris.

AtmoHEAD is a periodic 3-day international workshop on atmospheric monitoring and calibration for high-energy astroparticle detectors and experiments, with a view towards next-generation facilities. In total 24 talks (https://indico.cern.ch/event/536683/) were given this time in Olomouc covering wide field of various urgent topics from running experiments including the Pierre Auger Observatory, Telescope Array, H. E. S. S., MAGIC, HiGRO, IceAct. Emphasis was given also to plans for atmospheric monitoring of future observatories like CTA and EUSO as well as to discussions addressing common needs of the whole community. Talks were given by prominent experts from 8 countries and 3 continents all over the world.

The organizers of the event wish to thank to all the speakers and to all authors of the proceedings.

The organization of ATMOHEAD 2016 was supported by the projects of the Ministry of Education, Youth and Sports of the Czech Republic (LG14019, LE13012 and LM2015046), by the Palacky University in Olomouc and by the Institute of Physics of the Czech Academy of Sciences.

Conference photo by Dušan Mandát

The importance of atmospheric monitoring at the Pierre Auger Observatory

Bruce R. Dawson[1,*] for the Pierre Auger Collaboration[2,**,***]

[1] *Department of Physics, University of Adelaide, Adelaide, S.A. 5005, Australia*
[2] *Observatorio Pierre Auger, Av. San Martín Norte 304, 5613 Malargüe, Argentina*

Abstract. The Pierre Auger Observatory is an ultra-high energy cosmic ray experiment employing a giant surface array of particle detectors together with telescopes to image fluorescence light from extensive air showers in the atmosphere. The atmosphere is the medium in which the incoming cosmic rays deposit their energy, and as a result we must monitor the characteristics of the atmosphere, including its density profile and light transmission properties, over the Observatory area of 3000 square kilometres.

1 Introduction

The challenge of understanding the origin of the highest energy cosmic rays requires an observatory with a huge area and analysis methods which minimise uncertainties in cosmic ray energy, arrival direction and mass. The Pierre Auger Observatory, located near the town of Malargüe in Mendoza Province in western Argentina, covers a ground area of 3000 square kilometres [1]. It measures the extensive air showers initiated by ultra-energetic cosmic rays with energies from around 10^{17} eV to beyond 10^{20} eV using an array of over 1600 water-Cherenkov detectors, each of $10\,\mathrm{m}^2$ area and spaced on 1500 m triangular grid. During clear nights the air showers are also observed with 27 telescopes arranged in four sites around the edge of the surface detector array (figure 1). The telescopes detect the faint fluorescence light emitted by atmospheric nitrogen molecules excited by the passage of the air shower. The Observatory is thus a "hybrid" combination of two complementary techniques embodied in the Surface Detector (SD) and the Fluorescence Detector (FD).

The atmosphere is the medium in which the cosmic ray deposits its energy, and all Auger measurements by the SD and FD are influenced in some way by its properties. For example, the atmospheric density profile determines the way in which the air shower evolves with height above ground, and how the secondary particles scatter via Coulomb interactions. These things determine the shower particle density at ground level, and partly determine the lateral distribution of those particles, which in turn affects the energy estimate of the SD, $S(1000)$, the signal measured 1000 m from the shower core. In the case of the FD, the light attenuation properties of the atmosphere must be well known so that corrections can be made for an air shower that might be viewed up to 30 km from a detector.

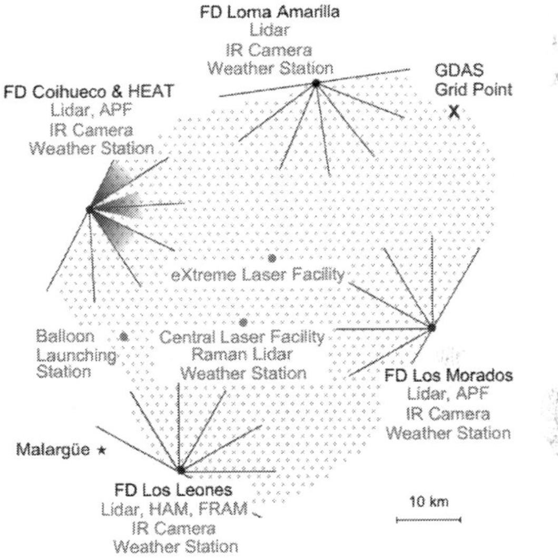

Figure 1. The Pierre Auger Observatory, showing locations of atmospheric monitoring instruments [1]. The surface detector array is indicated by the dots, and the fields of view of the fluorescence detector telescopes are indicated at each of the four FD sites.

The energy scale of the Observatory is determined using FD measurements. Fluorescence light is emitted from the shower in direct proportion to the energy deposited in the atmosphere, so the technique is intrinsically calorimetric, apart from a small correction for particles (neutrinos and high energy muons) that lose the majority of their energy in the Earth. Since the atmosphere imposes real and time-varying uncertainties on air shower measurements, Auger employs a number of atmospheric monitoring devices for characterising the properties of the atmosphere on time-scales as small as 5 minutes.

*e-mail: bruce.dawson@adelaide.edu.au
**e-mail: auger_spokespersons@fnal.gov
***Author list: http://www.auger.org/archive/authors_2016_09.html

2 Atmospheric Monitoring at the Observatory

The atmospheric monitoring instruments in place at the Observatory are described in [1] and [2], and their locations are indicated in figure 1. Ground level conditions (air pressure, temperature, wind-speed, humidity etc.) are monitored every five minutes by a series of weather stations at each of the four fluorescence detector sites, and towards the center of the array at the Central Laser Facility. For more than 6 years from 2002, regular meteorological radiosonde flights were launched from the Balloon Launch Station of the Observatory to characterize the vertical atmospheric profiles of pressure, temperature and humidity, and their seasonal variations. Two laser stations, at the Central Laser Facility (CLF) [3] and the eXtreme Laser Facility (XLF), have been in operation since 2003 and 2008, respectively. They fire mJ-scale pulses of 355 nm light vertically into the atmosphere which are detected by the FDs in order to measure the aerosol content of the atmosphere hour by hour during dark periods. The CLF and XLF have been the main source of our aerosol information, but they have been augmented by elastic lidar stations at each of the FD sites [2]. Two aerosol phase-function (APF) monitors are installed at the Los Morados and Coihueco FD sites. These collimated xenon flash-bulb light sources are used to measure the angular scattering characteristics of the aerosols close to ground level [2, 4]. Optical telescopes called HAM and FRAM are used to measure the wavelength dependence of the aerosol scattering, while the FRAM is also used to examine the sky in the directions of significant air shower events to search for unusual absorption by clouds or aerosol layers [2, 5]. Infra-red cameras mounted on pan-and-tilt platforms scan the night sky for cloud, and are used to fill a database detailing the cloud obscuration in the direction of every FD pixel every 5 minutes during FD operations [6]. Finally, since November 2013 a Raman lidar system has been in operation at the Central Laser Facility. Apart from the elastic scattering receiver for the 355 nm laser light, it has channels for Raman scattering off nitrogen and water molecules [7].

Information on the vertical density, temperature and humidity profiles of the atmosphere are sourced from the Global Data Assimilation System (GDAS) described in Section 4.2 below.

3 Event Reconstruction

An overview of event reconstruction at the Auger Observatory is given in [1]. Here we provide an introduction to the features most relevant to the present topic, atmospheric monitoring.

3.1 Surface Detector Reconstruction

The primary energy of the cosmic ray is estimated with the SD using the signal 1000 m from the shower core, $S(1000)$, where the distance is measured in a plane perpendicular to the shower axis. This is determined by fitting a lateral distribution function to station signals at various distances from the shower core. Showers of a given primary energy will have a different $S(1000)$ at different zenith angles because of the attenuation of the shower size at atmospheric depths past shower maximum. This is accounted for, on average, using the method of constant intensity cuts [1], which corrects $S(1000)$ to a value it would have if the shower had arrived at a zenith angle of 38°, the median zenith angle of air showers detected by Auger in the zenith range $0 - 60°$. The relationship between this corrected measure of the shower size at ground, called S_{38}, and the primary energy E is determined using hybrid events. The near-calorimetric measurement of E by the FD calibrates the S_{38} scale for use 24-hours per day.

Two subtle effects can affect the measurement of $S(1000)$ for showers of a given energy and zenith angle. One, related to variations in pressure away from the average value, will change the atmospheric overburden and induce small changes in the shower attenuation and hence $S(1000)$. The other effect, more related to changes in the air density in the lower ~ 1 km of the atmosphere, affects the Coulomb scattering of the electromagnetic portion of the air shower, and the shape of the lateral distribution function. These are small effects, contributing less than a 1% variation to the estimate of primary energy. However, when searching for weak, large scale anisotropies in the flux of cosmic rays, correcting for these effects is necessary. Such studies measure event rates above a fixed energy threshold. If that threshold energy is affected by weather effects, even small changes can result in measureable apparent anisotropies, given the steepness of the cosmic ray energy spectrum.

The collaboration has developed corrections to $S(1000)$ based on the air pressure and air density around the time of the event [8], determined using measurements at the weather stations described in Section 2. While not yet implemented in final data products like the energy spectrum (since the $< 1\%$ effect is small compared with the total systematic uncertainty on the energy scale of 14%), it is being employed in anisotropy studies such as [9].

3.2 Fluorescence Detector Reconstruction

The passage of an air shower through the atmosphere generates both fluorescence light (through the excitation of molecular nitrogen) and Cherenkov light. The Cherenkov light is much more intense, but is generally directed along the shower axis in contrast with the weaker but isotropically emitted fluorescence light. Significant amounts of Cherenkov light may be detected by the FDs if the shower axis points towards a telescope, or more commonly, when atmospheric constituents (molecules, aerosols or cloud) scatter part of the intense beam towards the FD.

The optical bandpass of the FDs is restricted to the range $300 - 400$ nm to match the spectrum of nitrogen fluorescence and to minimize the contribution of night-sky background. Within this band, absorption by atmo-

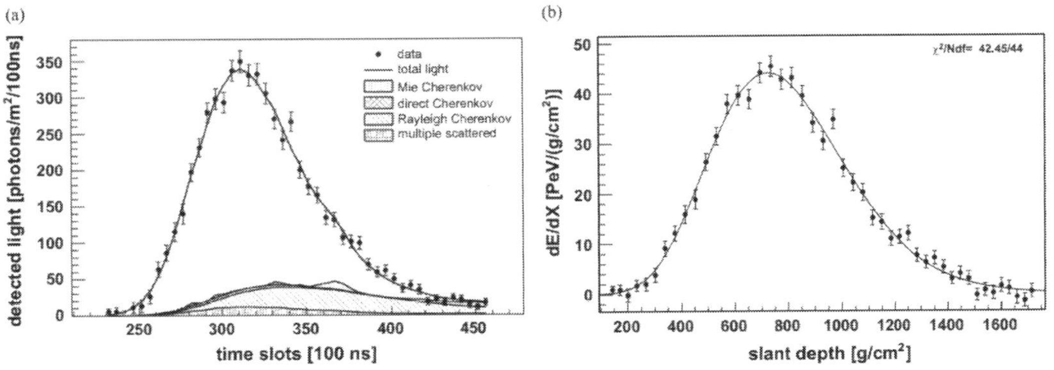

Figure 2. An example of a reconstructed shower profile. (a) The light detected at the telescope, including the fractions identified as being due to direct and scattered Cherenkov light. (b) After taking account of the shower geometry and light attenuation, application of the fluorescence efficiency gives the energy deposited along the shower track, dE/dX as a function of atmospheric depth [1].

spheric molecules is essentially zero, and light attenuation is due to scattering off molecules (Rayleigh scattering) and aerosols. One task of atmospheric monitoring is thus to understand the distribution of molecules and aerosols to account for (a) Cherenkov light scattered towards the detector, and (b) attenuation of light between the shower and the detector.

Once the detected signal is decomposed into fluorescence and Cherenkov light parts (the latter is typically less than 10%), knowledge of the atmospheric transmission will give the fluorescence light emitted from points along the observed shower track. Since the energy emitted as fluorescence light is a known fraction of the total energy deposited by the shower, our measurement yields the shower dE/dX as a function of atmospheric depth. Integrating this over depth (using a fit of the Gaisser-Hillas function, see [1] section 10.5) provides a measurement of the calorimetric energy E_{cal}. The primary cosmic ray energy estimate is E_{cal} plus a small ($\sim$ 10%) correction for "invisible energy", that energy carried into the ground by high energy muons and neutrinos. The atmospheric depth at which the energy deposit maximises, X_{max} is also measured, an important parameter for mass composition studies. An example of a reconstructed shower profile is given in figure 2.

The atmosphere and its measurement imposes systematic uncertainties on both energy and X_{max}. For energy, the atmosphere contributes a systematic uncertainty of about 6% at the highest energies, within the total energy systematic of 14% [1]. For the depth of shower maximum, the systematic uncertainty on X_{max} at the highest energies is dominated by the atmosphere and amounts to $^{+10}_{-7}$ g/cm^2 [10].

4 Examples of the Influence of the Atmosphere on FD reconstruction

Here we provide a few examples of the use of atmospheric monitoring in the analysis of fluorescence detector measurements. While the list is not comprehensive, we hope that these examples will indicate the breadth of information used to improve the quality of our analysis.

4.1 Fluorescence light yield

The fluorescence light emission from air showers is quite weak, especially compared with the Cherenkov light production, but it has the key advantage of being emitted isotropically. Very roughly, 4 photons are emitted from excited nitrogen per metre of track of a charged shower particle (and approximately independent of height), or in energy terms, about 10^{-5} of the ionization energy loss of the air shower particles is emitted as light. The emission efficiency is approximately constant through the troposphere due to the competition between excitation of the nitrogen (larger when the air density is higher) and collisional de-excitation (also higher at higher density) [11]. Molecules excited by the air shower can be quenched through collisions with nitrogen, oxygen or water vapour. For precise expectations of the light yield, the excitation and quenching processes must be well understood. For quenching, important parameters include the temperature and the partial pressures of the quenching species.

Over the past decade great advances have been made in laboratory measurements of the process, particularly by the AIRFLY Collaboration [12]. Measurements have included the precise spectral shape and intensity of emission between 300 and 400 nm, and the pressure, temperature and humidity dependence for each vibrational band within the spectrum. The temperature variation includes the expected dependence on $\sqrt{T}$ related to the probability of collisions, but also includes the more complex dependence of the quenching cross-sections on temperature.

In air shower analysis we make use of the GDAS three-hourly atmospheric profiles of pressure, temperature and humidity versus height (see next section). Together with the AIRFLY measurements, we are able to predict the absolute fluorescence yield with a systematic uncertainty of only 3.4% in the 337 nm band under standard atmospheric conditions, with an additional 1.1% systematic related to

DOI: 10.1051/epjconf/201714401001

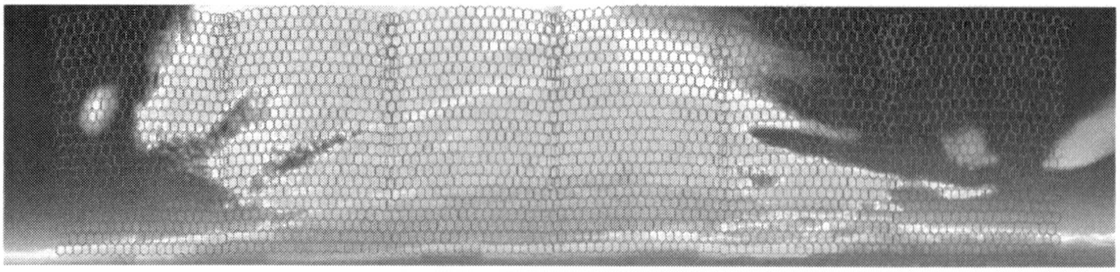

Figure 3. An example of an infra-red cloud camera image of the field of view of a fluorescence detector site. The warmer cloud is clearly visible against the colder clear night sky. Overlaid is the pattern of FD pixel fields of view for the six telescopes at this site, covering 180° in azimuth and approximately 30° in elevation. The database used in shower analysis contains an index indicating cloud coverage for every FD pixel, updated every 5 minutes during the night.

the spectrum across the entire band and our ability to account for the quenching [1].

4.2 Vertical Profiles of Density, Temperature and Water Vapour

Knowledge of the atmospheric conditions as a function of height within the troposphere is required for a number of analysis tasks. We have already described their use in accounting for the fluorescence yield. The vertical density profile is also used to convert height of a point on the shower track (from the geometrical reconstruction of the shower) into slant depth X (in g/cm^2), and in the calculation of Rayleigh scattering transmission from the shower to the detector.

For a period of six years from 2002-2008 the Auger Observatory undertook a comprehensive campaign of radiosonde balloon launches at the site to characterise the seasonal dependence of the vertical atmospheric profiles, resulting in local monthly models of air density, temperature and humidity. More recently we have switched to using data from the Global Data Assimilation System (GDAS), available since the beginning of 2005. GDAS provides information on the density, temperature and humidity profiles every three hours at points on the globe separated by 1 degree in latitude and longitude, including a point near the north-east boundary of the Auger Observatory, see figure 1. Auger has used its long-term radiosonde measurements to validate the coincident GDAS profiles [13]. Typically there is agreement to within 1 degree in temperature, 0.5 hPa in pressure, and 0.3 hPa in water vapour pressure at all relevant heights. GDAS thus provides profiles of sufficient accuracy every three hours, much more frequently than would be possible with any local radiosonde campaign.

4.3 Cloud Detection

The transmission of fluorescence and Cherenkov light from the shower to the detector can be affected by molecular and aerosol scattering, and by cloud. The effect of ozone absorption below 300 nm is negligible because of the FD optical bandpass filters have essentially no transmission at the relevant wavelengths. We have described above how GDAS density profiles are used to determine Rayleigh attenuation, and the next section will detail our aerosol measurements. Here we outline the several ways cloud is monitored.

The CLF and XLF laser facilities are the primary instruments for aerosol measurements but a by-product of that analysis is the minimum cloud-base height every hour. In addition, scanning elastic lidar stations at each of the four FD sites have been used to provide cloud height and cloud optical depth in a 45° cone in the vicinity of those sites on an hourly basis. Data from both the CLF/XLF and elastic lidars are accessed during shower analysis via calls to custom MySQL databases.

Complementary information is provided by scanning infra-red cameras, again one per FD site, which image the FD telescope fields of view (FOV) every 5 minutes and the entire sky every 15 minutes. In 2013 the cameras, having been in operation since 2002, were updated to new models with better stability and sensitivity [6]. The Gobi-384 uncooled microbolometer array cameras operate in the 8-14 μm range, and automatically scan the sky on pan-and-tilt platforms. They are radiometric cameras, meaning that they measure an absolute infra-red brightness temperature, allowing us to distinguish between warm cloud and cold clear night sky. The full-sky images are used for FD operator information, and the FOV images are processed for another MySQL database used in shower analysis, where a cloud cover index is listed for every FD pixel, updated every 5 minutes. Figure 3 shows an example FOV scan overlaid with the FD pixel pattern from one FD site.

We also access data from the GOES-12 and GOES-13 satellites for a broad overview of the cloud conditions above the Observatory [6]. A database is maintained with cloud probability maps derived from the satellite data twice per hour, with spatial resolution of 2.4 km by 5.5 km. The quality of this information has been validated by our ground based measurements [14].

Finally, the FRAM telescope scans the tracks of interesting air showers soon after observation to check for cloud and aerosols [5].

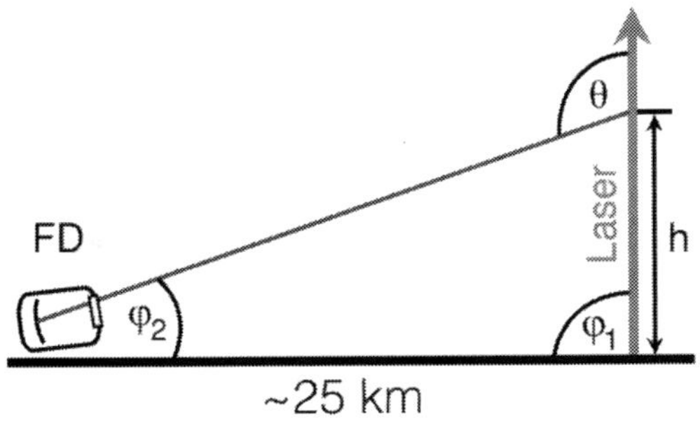

Figure 4. A sketch of the arrangement for CLF/XLF measurements of vertical aerosol optical depth, VAOD.

4.4 Aerosol Characterisation

Molecular, or Rayleigh, scattering is a more significant attenuation process than aerosol scattering. The molecular vertical optical depth between ground level and an altitude of 5 km is about 0.23 at 350 nm. This compares with an average value for the aerosol vertical optical depth of about 0.05 to the same height. However, the molecular atmosphere is much more stable than the aerosol one, and we have shown that the GDAS description of the molecular atmosphere is perfectly adequate. The challenge is to monitor the aerosol concentrations locally at the Observatory, and on time scales of an hour or less.

Our primary techniques use the Central Laser Facility (CLF) and the eXtreme Laser Facility (XLF) as part of a "bistatic" lidar for which the receiving optics are the FD stations (figure 4). In the past, cross-checks of the aerosol content have been provided by the standard elastic lidar stations at the FD sites. More recently, the FRAM telescope is contributing with aerosol estimates along the axes of interesting shower events [1, 7]. Finally, the "industry standard" technique, a Raman lidar, has been operating at the CLF site since November 2013.

The CLF and XLF lasers are frequency-tripled YAG lasers which shoot 6 mJ pulses of 355 nm light vertically into the sky [3]. (They are also steerable for other important tests, not described here). We use two approaches to measure the aerosol optical depth on an hourly basis using the CLF and XLF lasers. Both approaches make the assumption that there exists, from time to time, an aerosol-free atmosphere. We call this a "reference night".

The first approach, known as the Data Normalised (DN) method, makes direct use of hourly-averaged received light profiles on the reference night, and during the hour of interest. Examples are shown in the left-hand plot of figure 5. Each light profile is the average of 200 shots from a given hour (4 groups of 50 shots every 15 minutes), and normalised to a laser pulse energy of 1 mJ. The "bumps" in the profiles come from the imperfect light collection in the gaps between pixels in the FD camera.

When the laser pulse is shot vertically, portions of the light are scattered in the direction of the FD from all heights. The laser light suffers attenuation while travelling vertically above the laser, and in the path from the laser to the detector. By comparing the profile in a given hour to one from the reference night (where we assume that the only attenuation process is Rayleigh scattering) we can obtain (using an analytic expression [3], and taking care with possible cloud contamination) the vertical aerosol optical depth (VAOD) as a function of height, VAOD(h). An example is shown in the right-hand plot of figure 5.

This process is repeated for all four FD stations viewing either the CLF or the XLF (typically the XLF for the northerly Loma Amarilla detector), and the VAOD(h) data is loaded into a MySQL database for use during shower reconstruction. The VAODs derived within the same hour using different combinations of laser and detector site are consistent within uncertainties [3], implying near-uniform aerosol conditions across the Observatory at a given time. With the aid of the database of VAOD(h), and the horizontally-uniform aerosol assumption, the shower analysis can calculate the aerosol attenuation between any two points in space.

The DN method fills the majority of hours in the aerosol database. For hours where the DN fails for some reason, holes are filled with the alternate Laser Simulation (LS) analysis. As the name implies, we simulate the received signals from the CLF or XLF under a variety of aerosol conditions, and use this library of simulations to find the best match with a real hourly profile. For technical reasons (accounting for systematics in the laser simulation) a clean reference night is also used in this method. The DN and LS results for many hours have been compared and found to be consistent [3].

4.5 Cross-checks

In many ways the Auger Observatory's success rests on its demonstrated ability to cross-check results with different instruments or methods. Central to this philosophy is the hybrid nature of the entire Observatory, where (during dark periods) air showers are measured independently by the SD and the FD, detectors with completely different views of the showers and, largely, very different systematic uncertainties.

Much of this philosophy enters the atmospheric monitoring sphere, as may be obvious in the discussion above of various instruments used to measure cloud and aerosols. Partly, this is necessary to take advantage of the sensitivities of different instruments in different regions of phase space, but often cross-checks of the fidelity of data are also possible.

A recent example is the installation of a Raman lidar detection system using the CLF laser. The Raman technique allows an unambiguous solution to the lidar equation for aerosols by detecting inelastic backscatter from (in this case) nitrogen and water vapour [7]. The technique also has the advantage that there are standard quality assurance checks on such systems defined by experts in aerosol measurements around the world, leading to confidence in its results. On the other hand, using such a lidar in the clean aerosol environment of western Argentina is a very different situation to the majority of Raman systems located in urban settings. Work continues to validate the

 DOI: 10.1051/epjconf/201714401001

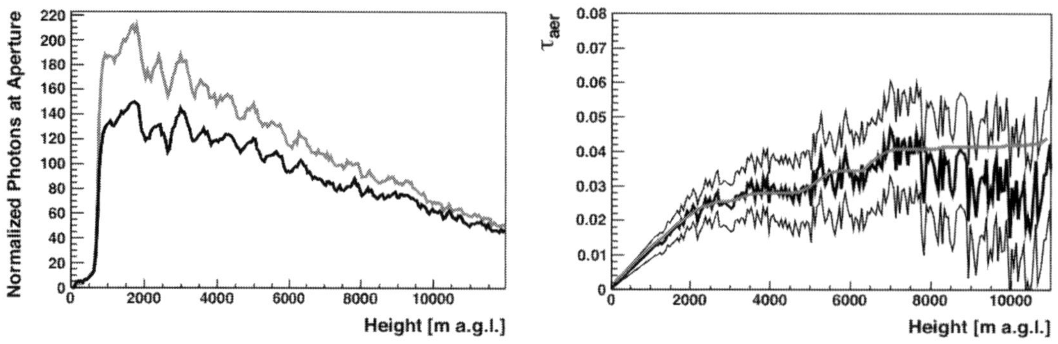

Figure 5. Left: laser profiles received at the fluorescence detector, normalised to a laser pulse energy of 1 mJ, and derived from 200 pulses over 1 hour. In red is the profile from the reference night, assumed to be aerosol free, with the black profile measured in the hour of interest. Right: the derived VAOD as a function of height above ground level, using the DN method. The central noisy line is the initial estimate of VAOD. A smoothed fit (red line) is also shown, derived with the constraint that the VAOD, being an integral quantity, cannot decrease with increasing height. Two other lines show the range of uncertainty in the VAOD [3].

standard Auger methods of aerosol characterisation using the Raman system, with three 12 minute Raman integrations per night (before and after nightly FD runs, and one integration in the middle of observations).

Finally, it is worth mentioning that shower data themselves offer scope for cross-checks of atmospheric measurements, particularly those related to aerosol transmission. Examples include comparing reconstructed energies for "stereo" views of air showers, those events seen by two, three or four FD sites, usually at different distances and aerosol transmissions. Similarly, even a single FD site energy estimate can be compared with the SD energy estimator to search for systematics related to aerosol transmission.

5 Conclusions

The monitoring of the important aspects of the atmosphere over the Pierre Auger Observatory's area of 3000 square kilometres is achieved with a suite of local instruments, and access to external data from radiosonde networks and satellites. A certain level of redundancy in the measurements is tolerated, and indeed welcomed for cross-checks. Creative uses of air-shower data has also been useful in understanding systematic issues in atmospheric monitoring. The Observatory is now embarking on a new upgrade, AugerPrime [15], with an emphasis on better SD measurements of mass-sensitive parameters, and an extension of FD operations into periods of increased night-sky brightness. Atmospheric monitoring will continue to have an important role in this new phase of the life of the Observatory.

References

[1] A. Aab et al. [The Pierre Auger Collaboration], Nucl. Instrum. Meth. **A798** 172–213 (2015).

[2] J. Abraham et al. [The Pierre Auger Collaboration], Astropart. Phys. **33** 108–129 (2010).

[3] P. Abreu et al. [The Pierre Auger Collaboration], JINST **8** P04009 (2013).

[4] S.Y. BenZvi et al., Astropart. Phys. **28** 312–320 (2007).

[5] J. Blazek et al., "Searching for Anomalous Shower Profiles with FRAM", and J. Ebr et al., "Aerosol measurements with FRAM telescopes", this meeting (2016).

[6] J. Chirinos et al., in "The Pierre Auger Observatory: Contributions to the 33rd International Cosmic Ray Conference (ICRC)", Rio de Janeiro, Brazil, 108–111 (2013) (arXiv:1307.5059).

[7] C. Medina et al., in "The Pierre Auger Observatory: Contributions to the 34th International Cosmic Ray Conference (ICRC 2015)", 127–130 (2015) (arXiv:1509.03732).

[8] J. Abraham et al. [The Pierre Auger Collaboration], Astropart. Phys. **32** 89–99 (2009).

[9] A. Aab et al. [The Pierre Auger Collaboration], Ap.J. **802** 111–121 (2015).

[10] A. Aab et al. [The Pierre Auger Collaboration], Phys. Rev. D **90** 122005 (2014).

[11] F. Kakimoto et al., Nucl. Instrum. Meth. **A372** 527–533 (1996).

[12] M. Ave et al. [AIRFLY Collaboration], Nucl. Instrum. Meth. **A597** 41–54 (2008); M. Ave et al. [AIRFLY Collaboration], Astropart. Phys. **42** 90–102 (2013).

[13] P. Abreu et al. [The Pierre Auger Collaboration], Astropart. Phys. **35** 591-607 (2012).

[14] P. Abreu et al. [The Pierre Auger Collaboration], Astropart. Phys.**50-52** 92–101 (2013).

[15] A. Aab et al. [The Pierre Auger Collaboration], "The Pierre Auger Observatory Upgrade AugerPrime" (2016) (arXiv:1604.03637).

Atmospheric Monitoring at the Site of the MAGIC Telescopes

Martin Will[1,⋆] for the MAGIC Collaboration

[1] *Instituto de Astrofísica Canarias (IAC), La Laguna, Tenerife, Spain*

Abstract. The MAGIC telescopes in La Palma, Canary Islands, measure the Cherenkov light emitted by gamma ray-induced extended air showers in the atmosphere. The good knowledge of the atmospheric parameters is important, both for the correct and safe operations of the telescopes, but also for subsequent data analysis. A weather station measures the state variables of the atmosphere, temperature, humidity and wind, an elastic LIDAR system and an infrared pyrometer determine the optical transmission of the atmosphere. Using an AllSky camera, the cloud cover can be estimated. The measured values are completed by data from global atmospheric models based on numeric weather forecasts.

1 Introduction

The Major Atmospheric Gamma Imaging Cherenkov (MAGIC) Telescopes are located in La Palma, Canary Islands, at the Observatorio de Roque de los Muchachos (ORM) at an altitude of about 2200 m above sea level. MAGIC is a stereoscopic system of two 17 m Imaging Air Cherenkov Telescopes. They measure very-high-energetic gamma rays (about 50 GeV to 50 TeV). The first telescope started taking data in 2004, a second telescope was inaugurated in 2009. Subsequent upgrades [1] made the two almost identical and improved their performance. A key feature of each telescope is the large mirror size of 235 m^2. MAGIC measures gamma rays with a sensitivity of about 0.66% of the Crab Nebula flux above 220 GeV [2], the energy and angular resolutions are between 15% and 24%, and 0.05° to 0.1°, respectively. Because of their light weight, a repositioning speed of 7°/s can be achieved to follow fast transients sources, e. g. Gamma-Ray Bursts.

1.1 The MAGIC Site

The ORM is located on the rim of the Caldera de Taburiente where several telescopes are placed close to the top at altitudes around 2400 m above sea level. The MAGIC site is at a lower altitude of 2200 m and generally shows a milder climate, better protected against strong winds from most directions due to the mountain slope. The observatory is situated far away from any significant industrial activities and protected by light pollution laws.

The atmosphere in the Canary Islands is characterized by its stability throughout the year. A temperature inversion layer usually appears around 1300 m a.s.l. on average, separating the moist marine boundary layer and the dry free troposphere. The ORM is always above this stable inversion layer with excellent observing conditions.

1.2 The Role of the Atmosphere for MAGIC

When a gamma ray interacts in the Earth's atmosphere, pair production and subsequent bremsstrahlung radiation produce secondary particles, initiating an avalanche process that generates hundreds of thousands of electrons and positrons. These secondary particles travel faster than light in the atmosphere, producing Cherenkov radiation. This emission is peaked in the ultraviolet, at a peak wavelength at ground around 290–300 nm.

The atmosphere affects the measured Cherenkov yield in three main ways: The development of particle showers depends on air density and refractive index; the Cherenkov light is absorbed between the emission point and the ground; and the Cherenkov light is scattered on the path to the detector, which causes a blurring and dimming of the detected images. The characterization and monitoring of the atmosphere is a very relevant part of the MAGIC data reconstruction. The main contribution to the systematic uncertainties of imaging Cherenkov telescopes stems from the uncertainty in the determination of the atmospheric transmission at a given time, wavelength, and altitude above ground. For MAGIC, 10% of the uncertainty of the energy scale is due to atmospheric effects [2], and 12% additional uncertainty are introduced on the flux due to run-by-run variations.

1.3 Safety Limits of the MAGIC telescope

The correct and safe operation of the telescope requires that weather conditions are appropriate. Within the safety limits, mechanical and electronic components of MAGIC work correctly and do not run any risk of being damaged. These components include the camera protection lids, the camera power supply, and the high voltage of the photomultiplier tubes. Additionally, the telescopes need be parked and secured in case of high wind or other severe

⋆e-mail: mwill@ifae.es

DOI: 10.1051/epjconf/201714401002

weather events. The following limits guarantee that the telescope hardware is not damaged:

1. Peak wind speed < 40 km/h (to open camera lids);
2. Average wind speed < 50 km/h (to move the telescope);
3. Relative humidity inside the camera < 60% (for high voltage);
4. Relative humidity outside < 90% (for data quality).

The part of the telescope that is most affected by wind are the lids of the camera. They extend on either side of the camera when opened and are closed automatically even for short wind gusts that exceed the limit. High wind speeds put great stress on all elements of the telescope dish structure, which can lead to positioning and pointing problems, or even bending or structural breaks. To avoid this, the telescopes are parked and secured if the average wind speed exceeds the limit. The high voltage of the camera is switched off in case of high humidity outside or inside the camera, as high humidity could result in short circuits. Both camera and power supply are designed to be water tight, and only the humidity inside the electronics is important. However, with the outside humidity approaching saturation, the high voltage is switched off as a safety precaution. In these conditions, no observations are possible anyway as the telescopes are in a cloud.

2 Atmospheric Monitoring Instruments

Several monitoring instruments are used at the MAGIC site. Most of them are located on the roof of the MAGIC Counting House in close proximity to the telescopes, see Fig. 1. These include a commercial weather station, an AllSky camera, and an elastic Lidar system. An infrared pyrometer is mounted on the dish structure of one of the telescopes.

2.1 Weather Station

The weather station used at MAGIC site is the `Reinhardt MWS-5MV` model. The temperature measurement is based on a precision platinum resistance thermometer in a range from $-40°C$ to $+60°C$ with an accuracy of $\pm$ $0.3°C$. The relative humidity sensor is a fast responding capacitive sensor based on a dielectric polymer, it can be used in a temperature range between $-40°C$ and $+60°C$. The range extends from 10% to 100%, with a measuring accuracy of $\pm$ 2%. The pressure sensor is a monolithic, laser-trimmed sensor, the measuring limits are 600 and 1100 hPa with $\pm$0.8 hPa accuracy. The anemometer measures wind speed using a Hall sensor, the wind peak and the average value are determined. The range is from 0 to 150 km/h with $\pm$ 2 km/h measuring accuracy. A weather vane with a precision magnetic encoder measures the wind direction in a range of 360°, the measuring accuracy is 5° above a starting speed of about 2 km/h.

The recorded data are transmitted to the serial port of a server machine in the MAGIC Counting House. Every two seconds, the station is read out and a data record including the time and date is received and written in a comma-separated text file. The Central Control and the Camera Control programs of MAGIC read this file and act in case a safety limit is violated. The data storage script reads the single string file every two minutes and generates a daily file in which the data are stored continuously. Long-term data is collected for more than 10 years at the MAGIC site.

2.2 AllSky Camera

An AllSky camera is installed to detect clouds. The aim is to identify those regions of the sky where the presence of clouds would discourage data taking. This information may be used to change the source schedule dynamically during the night. The camera used at the MAGIC site is the `SBIG AllSky-340` using the `Kodak KAI-340 CCD` with a resolution of 640 x 480 pixels. A fisheye lens with 1.4 mm focal length is mounted on top of the camera. The AllSky camera is mounted close to the weather station on the roof of the Counting House. A full picture is taken every 2 minutes, the images are stored as JPG images and in the FITS format. The pictures are then analyzed to extract information about cloud cover and star visibility. Comparing the found stars in the image and their magnitude with catalog values for position and brightness, patches of clear and cloudy sky are determined [3].

2.3 Elastic Lidar

A proper characterization of the atmosphere above the MAGIC site requires a real time range-resolved measurement of the transmission. The Lidar operated together with the MAGIC telescopes is a single-wavelength elastic Lidar. A specific Lidar inversion algorithm has been developed to obtain a vertical profile of the total extinction coefficient due to the presence of clouds and aerosols [4]. The Lidar at the MAGIC site is a low power system in order not to disturb the data taking of the MAGIC telescopes or the other telescopes at the ORM. An image of the Lidar can be seen in Fig. 1.

The system consists of a frequency-doubled, passively Q-switched, pulsed Nd:YAG laser. The wavelength is 532 nm, the pulse energy is 5 μJ, the pulse width is 0.5 ns. The pulse repetition frequency is up to 2 kHz, and the beam divergence is 0.05°. The telescope consists of a 60 cm aluminum mirror with a focal length of 1.5 m. The total weight of the mirror and the telescope frame is around 60 kg. The robotic, computer-controlled equatorial telescope mount `ASTELCO NTM-500` is used to move the Lidar. The maximum slewing speed is 30°/s, and the absolute positioning accuracy is 5 arcsec. A detector module using a Hybrid Photo Detector behind a selective band filter and optics is mounted in the focus of the mirror. The electronics for power supply and amplification are custom made. The charge of a single photo-electron is amplified in two stages. When a photon hits the GaAsP photo-cathode behind the entrance window, an electron is released with the probability of the quantum efficiency of about 50% at 532 nm. The electron is accelerated to the anode over a

DOI: 10.1051/epjconf/201714401002

Figure 1. The atmospheric monitoring instruments placed on the roof of the MAGIC Counting House (left). The small dome contains the elastic LIDAR system (right).

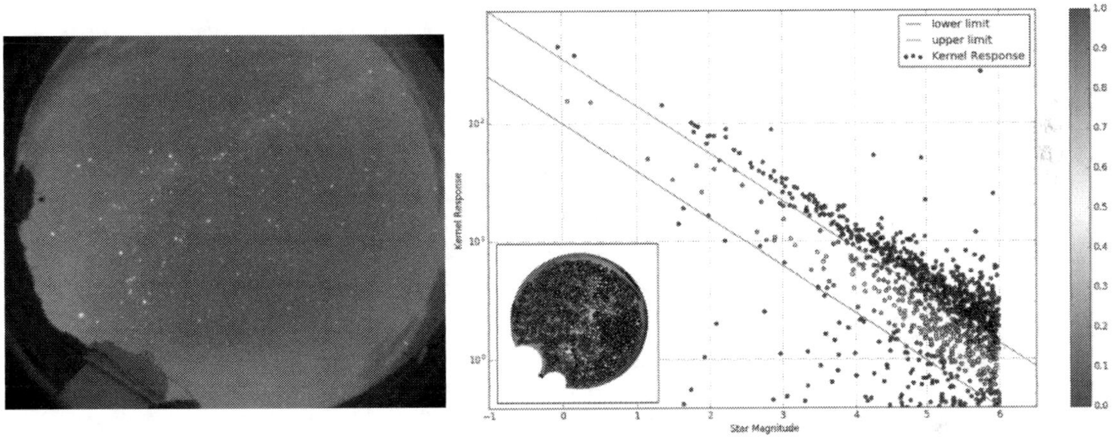

Figure 2. Image of the MAGIC AllSky camera, taken with 60 s exposure (left). By comparing the found stars and their apparent magnitude with star positions and magnitudes from a catalog, the level of cloud cover can be estimated [3].

several kV acceleration track. There it hits an avalanche diode that is operated in proportional mode. The 8 kV high voltage for the photon detector is supplied by a NIM module. The recording of the time profile of the backscattered photons is done using a computer equipped with a special FADC card.

Typically, the LIDAR tracks the current position of the MAGIC telescopes, a data run is taken automatically every 5 min consisting of 50 000 laser shots. The laser is fired along the optical axis of the LIDAR telescope, the light is reflected and scattered by molecules (Rayleigh scattering) and particles (aerosol scattering) in the atmosphere. The system offers a large dynamic range for a signal region ranging from distances of 0.5 to 18 km.

From the arrival time and the intesity of the return, the optical transmission can be calculated. This calculation is not trivial since two separate height-dependent parameters, the backscatter coefficient and the atmospheric extinction, have to be extracted from a single return. This is typically done by using complicated algorithms and several assumptions, but for the purposes of correcting the MAGIC data this effort is not necessary. The LIDAR returns are analyzed with two methods to determine the total extinction coefficient [4]. Both rely on relatively clean regions with a dominant Rayleigh scattering component before and after cloud or aerosol layers, and the excess in return signal due to additional scattering in-between, see Fig. 3. These procedures are possible because MAGIC observes in almost free tropospheric sky conditions.

The first method measures the total attenuation of a cloud layer by comparing the signal before and after the cloud, and using the excess over the Rayleigh scattering part of the signal, to extrapolate to the total aerosol volume scattering coefficient as a function of height. The second method uses an empirically determined LIDAR ratio (the extinction-to-backscatter ratio) of $K = 26.0 \pm 6.0$ for the typically thin clouds over La Palma to calculate the total aerosol volume scattering coefficient directly from the ex-

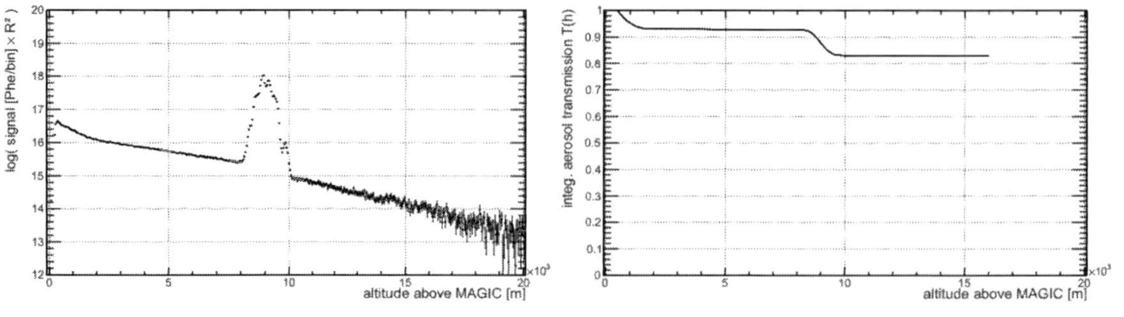

Figure 3. Analysis algorithm for analyzing Lidar data [4]. From the range-corrected signal (left), the integral atmospheric aerosol transmission is determined (right).

cess and the known total molecular scattering coefficient. The Lidar ratio was determined by using the extinction coefficient calculated with the first method and the backscattering coefficient from the Lidar signal for a selected sample of clouds. The vertical profile of the total extinction coefficient of the aerosol scattering is calculated. This profile can be converted into a cumulative transmission profile for the aerosol component, see right panel of Fig. 3, and will serve as input information for all further atmospheric corrections in the MAGIC data analysis chain.

For the analysis of the MAGIC data with corrections for variable atmospheric aerosol transmission, an approach is used which works well for low to medium aerosol extinction. The primary parameter that is affected by clouds or aerosols is the number of Cherenkov photons in the air shower images. The change in light intensity affects the energy reconstruction, which mainly depends on the size parameter of the Hillas parametrization of the recorded image. An air shower affected by aerosol extinction looks like an air shower of smaller energy. The correction can be done by scaling the size parameter to account for lower light content in the image due extinction. With a good approximation of the total light extinction, the energy estimation is up-scaled by the inverse weighted aerosol transmission of the atmosphere. In this way, the energy estimation of each event can be corrected using the real-time range-resolved information of the atmospheric aerosol scattering provided by the Lidar.

Using an optimized atmospheric calibration technique in the MAGIC data analysis chain enables a reliable use of data taken during moderate cloudy conditions. The effective duty cycle of the telescopes is extended by up to 15%. Some low energy events close to the threshold are lost, but for the higher energies, MAGIC gains significantly in observation time.

2.4 Infrared Pyrometer

A `Heitronics KT 19.82` radiation pyrometer is mounted on the side of the reflector surface of one of the MAGIC telescopes. It moves with the telescope and points to the same sky region as the telescope, the field of view of the pyrometer is 2°. It can be used to detect the presence of clouds in the MAGIC field of view. The pyrometer measures the integral thermal radiation from the sky in its line of sight using a thermoelectric sensor in the wavelength range from 8 to 14 μm. The measurement is passive and does not disturb the telescope operations.

The pyrometer measures the sky temperature above $-100°C$ with a resolution of $0.01°C$. The presence of a cloud can be infered since a cloud is "warmer" than the cloudless sky. The cloudiness C is a unitless value between 0 and 100 calculated using an empirically derived function of the sky temperature T_{sky}, zenith angle Z, air temperature T_{air} and relative humidity u,

$$C = 1.54 \cdot \left(T_{\text{sky}} - 0.31 \cdot T_{\text{air}} - 0.10 \cdot u - f(Z)\right).$$

Ground temperature and humidity data from the weather station on the roof of the Counting House are used in this calculation. $f(Z)$ is a function of the zenith angle.

An estimation of the cloud base height is also possible using the sky and ground temperature and assuming a constant temperature lapse rate $\Gamma = 6.49\,\text{K}\,\text{km}^{-1}$ in the lower atmosphere,

$$h_{\text{cloud}} = \left(T_{\text{air}} - T_{\text{sky}}\right) / \Gamma.$$

The influence of the humidity is not taken into account in this simple formula, a high water vapor content below the cloud changes the sky temperature.

The cloudiness and cloud base height as determined with the pyrometer are good additions to the transmission measurement from the elastic Lidar system and provide a good overall idea of the cloud cover above the site. As and integrated and not profile-resolved value, the cloudiness cannot be used to correct the gamma-ray data for the effects of clouds and aerosols as described above.

3 Model Data and Forecasts

MAGIC employs a collection of measurement instruments on site to determine the current atmospheric conditions. To increase the effectiveness of MAGIC observations, and to prepare for potential bad weather, model data and forecast systems are used.

3.1 Global Data Assimilation System (GDAS)

Data assimilation is used for the description and prediction of large-scale atmospheric variables, merging real-time measurements from a global, standardized network of instruments with numerical weather forecast models. The numerical models are based on non-linear equations, describing the global atmosphere with high precision on timescales of a few hours. Current data from weather stations, regular weather balloon launches, and satellite data are combined with predicted values from previous iterations of the assimilation process.

One such system is the Global Data Assimilation System. The data is available online, it covers the whole globe on a 1° by 1° grid in 3-hourly intervals. The grid point which is closest to the island of La Palma is considered, at 29° North, 18.00° West. The point is about 50 km north of the island above the sea, no grid point on the island itself is available. Since the MAGIC weather station is not part of the network feeding into the production of this model, a comparison between the two data sets is reasonable. In Fig. 4, the correlation between predicted GDAS temperature and the value measured by the MAGIC weather station is shown, separately for daytime and nighttime. The correlation is generally good, the small offsets in temperature can be explained by the fact that the GDAS grid point is above the sea as described earlier and residual heat from surrounding structures that influence the measurements. The small offset of 0.6 mbar in pressure lies within the uncertainties of the altitude of the MAGIC weather station (± 3 m) and the precision of the pressure sensor (± 0.8 hPa).

Since the GDAS model is valid at the MAGIC site, the profiles of the atmospheric state variables up to around 25 km can offer many more possibilities to use this information than the single data point the weather stations provides. One important aspect is the determination of the Rayleigh scattering in the atmosphere to improve the data analysis of the elastic Lidar system. When determining the aerosol transmission, the scattering of the returning laser light on air molecules has to be taken into account. Without knowing the detailed profile of the air density, only estimations of the Rayleigh scattering based on an exponential density profiles can be included. Using GDAS, the scattering can be calculated using the density profile at the moment of the Lidar data taking, which significantly improves the reconstruction of the transmission at higher altitudes.

3.2 Weather and Aerosol Forecast

To predict the conditions at the MAGIC site hours or even days in the future, several weather services provided online are used. Using the latest images processed from the EUMETSAT satellite data, the cloud conditions can be extrapolated in a simple way to get some idea of weather systems moving towards or away from the Canary islands. More sophisticated forecasts can be obtained through mountain weather forecasts[1] websites providing detailed information on temperature and precipitation developments in the upcoming week.

The SKIRON Dust System[2] is used to predict and forecast dust concentration, aerosol load and deposition in several sizes at the MAGIC site. The model resolution is 0.24° x 0.24° on 38 vertical levels from the surface up to 22 km, the temporal resolution is one hour, the forecast duration is 5 days.

4 Outlook and Conclusion

In the near future, MAGIC is planning to implement an adaptive scheduling system. Depending on atmospheric conditions, the schedule of observed sources is changed. This system will combine all the atmospheric data gathered at the site and available otherwise. The weather station data is used to guarantee the safety of the telescopes; the AllSky camera is used to map clear and cloudy sky patches; the elastic Lidar measures cloud height and thickness, and considering the GDAS model the transmission of the atmosphere is determined, complemented by data from the pyrometer; and satellite data and aerosol forecasts can be used to predict the progression of the conditions.

Using those information, the potential lower energy threshold of the MAGIC observations dependent on zenith and transmission can be determined. Only sources which are expected to be seen above this estimated threshold and in a clear part of the sky will be observed. This adaptive scheduling is currently in its final testing phase and will be implemented in the coming months. Potentially, 20% more data of analyzable quality can be gained through this way of scheduling sources.

Several instruments to measure atmospheric parameters and data from external sources is available at the MAGIC site. Measurements have been ongoing for more than one decade, guaranteeing the safety of the telescopes and the high-quality gamma-ray data MAGIC is known for. Correction of data taken under adverse conditions is possible and routinely done in the MAGIC data analysis, increasing even further the potential science output of the instruments. Even more analyzable data is expected through the use of adaptive scheduling taking into account the real-time information available.

Acknowledgments

The author would like to thank the Instituto de Astrofísica de Canarias for the excellent working conditions at the Observatorio del Roque de los Muchachos in La Palma. The financial support of the German BMBF and MPG, the Italian INFN and INAF, the Swiss National Fund SNF, the he ERDF under the Spanish MINECO (FPA2015-69818-P, FPA2012-36668, FPA2015-68278-P, FPA2015-69210-C6-2-R, FPA2015-69210-C6-4-R, FPA2015-69210-C6-6-R, AYA2013-47447-C3-1-P, AYA2015-71042-P, ESP2015-71662-C2-2-P, CSD2009-00064), and the Japanese JSPS and MEXT is gratefully acknowledged. This work was also supported by the Spanish Centro de Excelencia "Severo Ochoa" SEV-2012-0234 and SEV-2015-0548, and Unidad de Excelencia "María de Maeztu"

[1] http://www.mountain-forecast.com/

[2] http://forecast.uoa.gr/

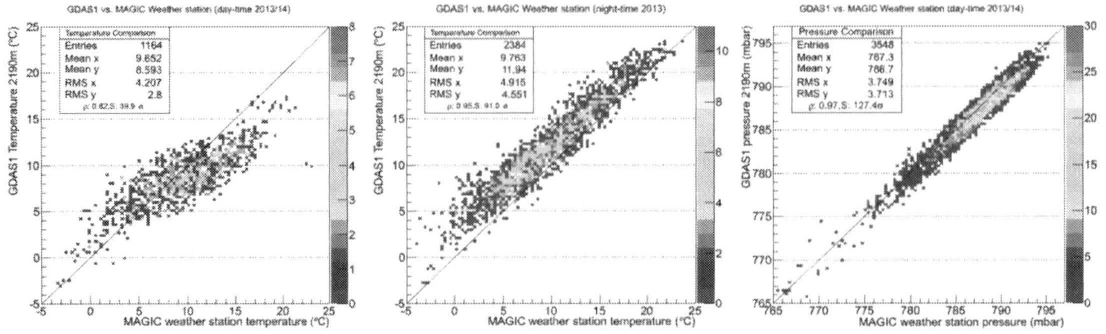

Figure 4. GDAS cross-correlation of atmospheric pressure (left), daytime temperatures (middle), and nighttime temperatures (right) with the measurements of the MAGIC weather station.

MDM-2014-0369, by grant 268740 of the Academy of Finland, by the Croatian Science Foundation (HrZZ) Project 09/176 and the University of Rijeka Project 13.12.1.3.02, by the DFG Collaborative Research Centers SFB823/C4 and SFB876/C3, and by the Polish MNiSzW grant 745/N-HESS-MAGIC/2010/0.

References

[1] Aleksić et al., Astropart. Phys. **72**, 61–75 (2016).
[2] Aleksić et al., Astropart. Phys. **72**, 76–94 (2016).
[3] J. Adam, *Cloud Detection and Prediction with All Sky Cameras*, these proceedings.
[4] Fruck et al., Proc. to the 33rd ICRC, Rio de Janeiro, PoS, 1237 (2015).

CTA Atmospheric Calibration

Markus Gaug[1,2,*]

[1] *Unitat de Física de les Radiacions, Departament de Física, Universitat Autònoma de Barcelona, Spain*
[2] *CERES-IEEC, Universitat Autònoma de Barcelona, Spain*

Abstract. The main contribution to the systematic uncertainties of Atmospheric Imaging Cherenkov Telescopes (IACTs) currently stems from the uncertainty in the determination of atmospheric properties for a given moment of observation. This technique uses the atmosphere as a calorimeter of gamma-ray induced atmospheric air showers and measures the amount of Cherenkov light collected by large mirrors and focused towards a pixellized camera. Atmospheric conditions affect the measured Cherenkov light yield in several ways: the air-shower development itself, the variation of the Cherenkov angle with altitude and the loss of photons due to scattering and absorption of Cherenkov light. Although IACTs use to be located at astronomical sites, characterized by extremely clear atmospheric conditions, the local atmosphere should be continuously monitored, in terms of molecular density profiles, aerosol extinction profiles, and clouds. Moreover, a general understanding of the site and particularly aerosol climatology is desirable for a robust interpretation of the sensing instruments' data. Here, the strategy of the Cherenkov Telescope Array (CTA) on atmospheric monitoring is presented, designed to ensure that the related systematic uncertainties are brought down by at least a factor of two with respect to current installations. Additionally, a more intelligent scheduling scheme is aimed at, where source observations are selected using dedicated auxiliary atmospheric monitoring instruments and taking into account the feasibility of a later analysis according to the underlying scientific case. This plan should reduce data loss during offline data selection and enhance the effective duty cycle of the CTA.

1 Introduction

Over the past decade, Imaging Atmospheric Cherenkov Telescopes (IACTs) have opened a new observation window for gamma-ray astronomy: the Very High Energy (VHE) regime, ranging from tens of GeV to several tens of TeV gamma-ray energy. The technique being ground-based permits building large mirrors and exploit the huge area illuminated on ground by the Cherenkov light emitted by gamma-ray induced air showers. This makes it possible to achieve unprecedented sensitivities, orders of magnitude better than what can be achieved by satellites in that energy range [1]. So far, more than hundred VHE gamma-ray sources have been discovered by IACTs [2].

On the other hand, using the atmosphere as a calorimeter brings along a series of limitations on the accuracy with which the energy of an air shower can be reconstructed, and the exposure of the instrument calibrated. Albeit several efforts have been made in that direction [3, 4] and unlike other astronomical installations operating at lower energy ranges, there is still no direct cross-calibration with satellites available, let alone a test beam.

The main contribution to the systematic uncertainties of IACTs stems then from the uncertainty in the (height- and wavelength-dependent) atmospheric transmission for a given moment in time. Current estimates range around 15% for the accuracy with which the energy scale is calibrated [5, 6] and a similar, or only little smaller number for exposure calibration. Both estimates are however based on selected data recorded during perfect atmospheric conditions and have to be considered lower limits for the general case of data taken under still acceptable conditions.

Atmospheric quality affects the measured Cherenkov yield in several ways [8]: the air-shower development itself depends on the molecular density profile along its trajectory as well as the variation of the Cherenkov angle. Further, the loss of Cherenkov photons is affected by absorbing molecules and, more importantly, scattering out of the camera field-of-view (FOV), both by molecular (Rayleigh) and aerosol (Mie) scattering. Clouds are usually thinner than the several kilometers long air showers and normally remove only a part of the emitted Cherenkov light from the FOV of an IACT. Since the mean interaction depth of the shower maximum depends on the primary gamma-ray energy, the altitude of a cloud layer can create energy dependent effects [9, 10]. Also the case of negligible interaction of Cherenkov light with a (high) cloud is possible, if the shower energy is high enough and even for an optically thick cloud [11]. An altitude-dependent assessment of the atmospheric extinction is therefore mandatory, e.g. with the use of a LIDAR [12]. Multiple scattering of photons back into the camera in turn, has only a minor effect and can normally be neglected [8].

Despite several auxiliary instruments of IACT installations are currently used to measure atmospheric extinc-

*e-mail: markus.gaug@uab.cat

tion [13], or even the very same science data can be exploited to that aim to certain extent [14, 15], their outcome has been mainly used to retain good-quality observation time slots, and only a minor effort has been made to correct data with atmospheric information [10, 13, 16]. Only the MAGIC collaboration has started to routinely include simultaneously taken LIDAR measurements in its data analyses [12, 17–19].

Following the natural evolution of new astronomical instruments and observation windows, the next generation VHE gamma-ray observatory Cherenkov Telescope Array (CTA) will move from the discovery mission of its predecessors to perform population studies and precision measurements, resolving spectral features and the morphology of gamma-ray sources with unprecedented precision [20][1]. A thorough calibration strategy for the used calorimeter, the air, must hence be included in its design, as well as ways to optimize its effective duty cycle, even after data selection and analysis cuts are applied [7]. Recently, the CTA site selection has been completed, with as outcome, for its Southern observatory (CTA-S), a plateau of the Cerro Armazones in Northern Chile, close to Paranal, while the "Roque de los Muchachos" Observatory (ORM) on the island of La Palma, Canary Islands, has been selected for its Northern part (CTA-N). Both observatories are located around 2200 m a.s.l. Knowledge of the concrete sites permits now to design tailored atmospheric calibration strategies for both.

This proceeding presents first several aspects of the atmosphere, necessary to understand CTA's atmospheric calibration strategy, followed by the designed calibration of the observed FOV. Later on, we will show how the effective duty cycle can be improved already during scheduling of sources.

2 The atmosphere above CTA

From the various atmospheric layers, only the troposphere (reaching from ground to ~15 km a.s.l.), and sometimes parts of the tropopause are affected by variations of their optical properties. Only in the case of stratovolcanic eruptions, the lower stratosphere (15–20 km a.s.l.) may become also optically degraded [24]. Air molecules can travel to the top of the troposphere and back down in a few days, hence the mixing of the characteristics of the layer is fast [25]. The lowest part of the troposphere, enclosing most of the aerosol over-density, is called the "planetary boundary layer" (PBL), whose thickness increases during day and shrinks during night. Its top is characterized by a sharp decrease in particulates and condensates frequently leading to the creation of convective clouds. During night, the PBL may leave a residual layer on its top, the "nocturnal boundary layer" where pollutants emitted at the surface remain concentrated in a layer few tens of meters thick near the surface. At an astronomical site, this layer is expected at about one kilometer above ground, however often showing an exponential decay of aerosol densities [26, 27].

While the molecular content of the atmosphere varies very slowly at a given location during the year, and slowly from place to place, aerosol concentrations can vary on time-scales of minutes and travel large, even intercontinental, distances [28]. Most of them are concentrated within the first 3 km of the troposphere, being the "free troposphere" above orders of magnitude cleaner. A "photometric night" at an astronomical site is practically identical with an undisturbed free troposphere, and a negligible amount of aerosols close to the ground. While ground-level aerosols (due to the nocturnal boundary layer, but also from desert dust intrusions or biomass burning) typically degrade the transmission of all Cherenkov light [9], the vertical structure of higher altitude aerosols and clouds (including cirrus) must be determined since only a part (or even none!) of the emitted Cherenkov light suffers extinction. A correction is then more difficult, but nevertheless possible [17]. Stratospheric aerosols from volcanic eruption debris in turn, present for years after strong stratovolcanic eruptions, may cause a significant obscuration of the star light [24], but only influence the Cherenkov light from air showers of exceptionally high shower production heights and hence only at the very energy threshold of an IACT. Relying only on photometry of reference stars may introduce hence a strong bias difficult to correct for [8].

The understanding of the evolution of the main characteristics of aerosol (size, absorption and scattering of light, chemical composition, etc.) is extremely complex and a prime topic of current research, given their importance for the understanding of the world climate [29]. Light scattering and absorption by aerosols needs to be described by Mie theory or further developments of it, including non-sphericity of the scatterer [30]. Aerosols use to have larger refraction indices than the one of water, and typically show also a small imaginary part [31, 32]. Contrary to the typical λ^{-4} wavelength dependency of Rayleigh-scattering molecules, aerosols show power-law indices (the so-called Ångström coefficients) from 0 to 1.5, i.e. a much weaker dependency on wavelength [33, 35]. Of special interest here is the ratio between extinction and backscatter coefficient, the "Lidar Ratio" (LR) which shows a strong dependency on relative humidity and can range from 15 to 80 sr [33] for ground layer aerosols, and from 2 to 20 sr in cirrus clouds [34] (the pure Rayleigh scattering case corresponds to $8\pi/3$ ~8.4 sr). Elastic LIDARs need to make a-priori assumptions on this parameter, their main systematic uncertainty stems from the conversion of the (measured) backscatter coefficients to the (inferred) light extinction coefficients, as a function of height. Raman LIDARs in turn, are able to measure the LR directly.

3 Overall Strategy

Following the CTA's science and performance requirements, the allowed systematic uncertainty on the reconstructed overall energy scale of CTA permits some 7–8% maximally as contribution from atmospheric calibration. This is in fact the uncertainty with which the Cherenkov

[1]Several recent discoveries by current IACTs are already limited by the systematic uncertainties with which the instrument response is known. What was good enough for the discovery of sources, is now starting to limit performance [see e.g. 21–23].

DOI: 10.1051/epjconf/201714401003

photon density on ground is related to the air shower energy on average. The relation is estimated via comparison of the measured number of Cherenkov photons with its prediction from Monte Carlo (MC) simulations of gamma-ray induced showers. Although a given fixed atmosphere must be used for each simulation set, the CTA will aim for time-interval (TI)-wise simulations, configured each to match the given state of hardware and the respective atmospheric conditions.

Since Cherenkov photons can be emitted (and imaged) anywhere from about 20 km altitude to ground, the troposphere and lower stratosphere need to be characterized in a height-dependent way.

3.1 The molecular profile

A first topic of prime interest concerns the need (or not) to adapt molecular profiles for each simulation TI. Although initial work by Bernlöhr [8] pointed to principally possible large variations, of the order of up to 20% in Cherenkov light yield for mid-latitudes, the actual conditions at the locations of the H.E.S.S. and MAGIC telescopes [8, 36] resulted to be such that one average profile was sufficient as input for all MC simulations. The assumed error of only 2–3% rms induced by this choice resulted to never limit their overall systematic uncertainty, by a factor of several times higher [5, 6]. Whether such a statement is still true for the CTA, is currently being investigated [37, 38]. The difference is not only due to the stricter requirements on the CTA's accuracy, but also to its wider coverage of gamma-ray energies. Towards the high energy range (aiming at energies larger than 300 TeV in the South) showers penetrate deeper into the atmosphere which in turn shows different density variations [see e.g. 37]. On the other side of the spectrum, very low energy showers develop in an around the tropopause where relative density variations, and their uncertainties, are largest [37].

Variations of the Cherenkov light yield are due to the differences in the height of the shower maximum, coupled with the height-dependent threshold for Cherenkov light emission [8, 39], and the modulation of the Cherenkov angle with the local density of air. If measurements or now-cast predictions with a precision of about 1 g/cm^2 at a height resolution of 20 m are available, about 2.5% uncertainty on the relative amount of radiation lengths of electrons and gamma rays can be achieved, and hence approximately a relative uncertainty on the shower energy of a same order of magnitude. Modern radio sondes can easily achieve such a precision, albeit diurnal and seasonal variations may be greater [27, 37]. Apart from air shower physics, the IACT measurement process is moreover determined by the production mechanism of Cherenkov light which itself depends on the refractive index (and hence more generally the density) of the surrounding air at its emission point. When the Cherenkov cone opening angle increases, the Cherenkov light gets spread over a larger area, decreasing the central light density. Bernlöhr [8] has shown that the central light density ρ_c for vertical showers follows roughly:

$$\rho_c \propto \left(h^*_{\mathrm{med}} - h_{\mathrm{obs}}\right)^{-2} \quad , \qquad (1)$$

where h^*_{med} is the median height of emission for Cherenkov light near the core (typically 1000–1500 m below that of all Cherenkov light and hence the shower maximum) and h_{obs} is the observatory altitude. From Eq. 1 follows immediately that any constant relative systematic offset of the molecular altitude profile is expected to create stronger relative deviations in Cherenkov light density for high shower energies (i.e. smaller values of h^*_{med}). Studies carried out for MAGIC, using the now-cast predictions of the "Global Data Assimilation System" (GDAS)[2] for La Palma, have shown that a constant difference of 1% between the assumed and the true integrated density from ground to the shower maximum (corresponding to the rms of the GDAS predictions throughout one year) leads to differences in ρ_c of about 0.5%, 1.5%, 3.5% and 6% for gamma-ray energies of 20 GeV, 200 GeV, 3 TeV and 70 TeV, respectively applying Eq. 1. To these numbers, an additional contribution of the probable bias between the mean GDAS model and the true average annual atmosphere needs to be added, which must be validated by dedicated radio sonde campaigns [37], or a LIDAR sufficiently sensitive to the molecular profile [27], or from sophisticated analysis of high-resolution spectrometer data gathered by an optical telescope [40]. The CTA will carry out a radio sonde campaign at its Southern site and rely on a database covering several years of MAGIC LIDAR [12] data to assess the latter. A dedicated simulation effort is now ongoing in CTA [38] to directly assess the effect of typical and maximum variations of the density profile using the vast historical database of the GDAS[3], together with assumptions on the accuracy of these. The outcome of these simulations should be ready for the next AtmoHEAD conference and answer the question of how many different molecular profiles will be required for the CTA data analysis.

Light absorbing molecules are of less importance for the CTA, given the drop of Cherenkov light intensity with increasing wavelength, and consequently the chosen sensitivity window of the telescope cameras, limited to between 290 nm and about 700 nm. In this range, only few molecules (mainly O_3, NO_2, H_2O and O_2) absorb light at all, of which ozone creates the biggest effect, both in the Huggins band below 360 nm as well as the weak Chappuis band (375–650 nm). The strong Hartley band (220–310 nm) reaches only marginally into the sensitive wavelength region of the CTA. The strongest systematic effect is expected from ozone intrusions into the upper troposphere, so-called "stratosphere-troposphere exchange" (STE) processes, best assessed by radio sondes with ozone measuring capabilities [41]. Since these come along with a considerably increased price adding to the already high cost of a thoroughly and systematically designed radio sonde campaign, cheaper methods have been investigated. For the North, and particularly the Canary Islands, Cuevas et al. have established an ozone climatology [42] using the "Potential Vorticity" (PV) parameter [43] obtained from "European Centre for Medium-Range Weather Forecasts"

[2]https://www.ncdc.noaa.gov/data-access/model-data/model-datasets/global-data-assimilation-system-gdas, see also http://earth.nullschool.net/

[3]or, alternatively the ECMWF http://www.ecmwf.int/

DOI: 10.1051/epjconf/201714401003

(ECMWF)[4] reanalysis, and compared with backward trajectories using the "Hybrid Single-Particle Lagrangian Integrated Trajectory" (HYSPLIT)[5] code. They show that the Canary Islands are basically un-affected by STE processes throughout the year, a finding confirmed by Škerlak et al. [46]. However, the region of Mid-Northern Chile seems to be hot spot for STE [46], at least during the austral Summer months, centered about 10° south of CTA-S. Rapid increases, on time scales of less than a day, in ozone mixing ratios linked with STE processes [44] have been observed at Cierro Tololo, a pristine mountain site about 5° south of CTA-S. Strong deep stratospheric mass flux into the PBL was also observed by Škerlak et al. [46] during the austral winter months over the Andes, from 20–30° S. An analysis of 20 years data of the same site [45] revealed an overall positive trend of ~0.7 ppb/decade in surface ozone, and strong seasonal trends, with maxima in October, again correlated with STE processes and during La Niña conditions. Generally, the contribution of ozone exchange with the stratosphere is predicted to increase with climate change [47]. Kimeswenger et al. [48] however suggested to build a specialized spectrograph mounted on an optical telescope to monitor the ozone profile. Whether ozone monitoring with a Cimel Sun photometer[6] is feasible for the CTA-S still needs to be found out. The loss of Cherenkov light due to ozone is however expected to be less than 5%, with an uncertainty of a few percent only.

3.2 Aerosol characterization of the observed field-of-view by CTA

The main contribution to systematic uncertainties in atmospheric calibration stems from aerosols and clouds which can show variability over minutes only. Consequently, aerosol extinction needs to be monitored on such time scales. Since some interpolation in time is always necessary and adds additional uncertainties, the combined precision and accuracy of each aerosol optical depth (AOD) measurement, as a function of emission altitude, should not exceed 2%, with an altitude resolution of the order of one electron radiation length, i.e. ~40 g/cm^2. Only the fine-structure of the nocturnal boundary layer need not be resolved [9]. Moreover, optically thick low clouds are of no interest since observations will be aborted anyhow under such conditions. Remain optically thin clouds, at any altitude, which are typically thinner or of the order of one radiation length. Hence most of the time, even the altitude resolved fine-structure of clouds are of no interest, only their integral optical depth.

The CTA has been designed to cover a wide FOV, reaching up to 10° per telescope, and up to 20° for the full array, if operated in divergent pointing mode. Such large FOVs entail the necessity to characterize a similarly large patch of the atmosphere, the characterization in only one line-of-sight results insufficient.

Raman LIDAR measurements are desirable to determine the altitude profile of aerosol extinction since they provide the required precision and accuracy for both ground-level aerosols (if equipped with near-range optics or if the LIDAR's system constant can be calibrated [see 27]) and the vertical structure of higher altitude aerosols and clouds.

Experience has shown that Raman LIDARs and their data analysis methods should be validated through recognized LIDAR networks, e.g. the EARLINET[7] in Europe. We plan to install one Raman LIDAR in the Northern observatory and two Raman LIDARs in the South, all of them validated directly or indirectly (e.g. through the ARCADE project [49, 50]) by such a LIDAR network. All proposed LIDARs will be operating at 355 nm and 532 nm at least, in order to determine the Ångström coefficient in the wavelength range where the CTA telescopes are sensitive to Cherenkov light.

Detailed calculations of the sensitivity of the proposed Raman LIDAR projects have been carried out [51, 52], developed with the aim to reach high precision in the derivation of the aerosol extinction until high altitudes, without introducing observation time loss to the observatory. In this study, the dimmest Raman line will be detected from a distance of 13 km (i.e. altitude about 15 km a.s.l., if observed at zenith) with a signal-to-noise ratio of 10 in only one minute. This will permit to fully characterize one line-of-sight a few minutes before and after a CTA observation run, and during the change of Wobble position [53].

Nevertheless, LIDARs cannot scan the FOV observed by the CTA and continuously disturb science observations. Instead, they will be accompanied by a wide-angle telescope, the FRAM [54, 55], designed to produce atmospheric extinction maps through various standard optical filters with high angular resolution and covering a FOV of about 15°×15°. Maps of integral AOD, each patch determined with better than 3% precision (see [55, 56] for an explanation of the underlying algorithms), are then taken several times during a CTA observation run and interpolated in time. Interlaced Raman LIDAR measurements serve to divide the interpolated AODs in vertical bins and to re-calibrate its integral. These "aerosol extinction hypercubes" (with altitude, wavelength and time as additional dimensions) are then split into a slow component (due to molecular extinction and quasi-stable aerosol layers, like the boundary layer or sometimes even clouds), and a fast one, which changes throughout a science observation run. Historical information from aerosol characterization campaigns [49], and later on the very same database of previous LIDAR/FRAM observations, together with now-cast predictions of the molecular profile from the GDAS or similar, will help to produce the extinction hypercubes and the identification of their components. The expected shifts in CTA performance from a given extinction hypercube are continuously confronted with those of currently used (or foreseen) MC generated Instrument Response Functions (IRF), and a series of systematic errors is calculated and confronted with the CTA requirements. In case one of these systematic errors exceeds the allowed limit, a new "Time Interval" (TI) is communicated to the

[4]www.ecmwf.int/
[5]www.arl.noaa.gov/HYSPLIT.php
[6]www.cimel.fr
[7]https://www.earlinet.org/

CTA Software, together with a suitable start time, and an averaged extinction hypercube, to be used for a new MC simulation of the atmosphere. The critical part of this procedure is a robust estimate of the systematic error. For this purpose, MC simulation studies are required to determine the impact of aerosols and clouds at different altitudes, and cirrus clouds covering parts of the FOV of a Cherenkov camera on the angular and energy resolution and bias, if a uniform coverage of the CTA cameras by aerosols has been assumed in the simulations. Previous studies of the typical morphology of clouds and evolution time scales at the CTA sites are needed as input for these simulations [57].

Additionally to that, redundant information is obtained for calibration and cross-checks by photometric measurements of reference stars using a sophisticated, portable small-aperture multi-pixels photon detector, the UVScope and the UVSiPM [58, 59] and the "Cherenkov Transparency Coefficient" (CTC) calculated directly from science data [60].

4 Intelligent target selection

Current IACTs effectively lose a considerable amount of their effective duty cycle at the moment when data selection is made offline. This is due to the fact that observation targets use to get chosen according to a fixed schedule and regardless of atmospheric conditions (as long as observations are not aborted at all). The observations suffer then from non-optimal atmospheric conditions along their line-of-sight, instead of actively avoiding them. However, optimal atmospheric conditions may be decisive for the practicability of low-energy analyses [61], those requiring precision pointing [62], or those requiring maximum sensitivity for monitoring purposes, while other types of observations may be practically unaffected. Selecting data according to these criteria later on offline is hence the worst of all strategies, at least in terms of achievable duty cycle, and if sources with less sensitivity to such non-optimal conditions could have been observed instead.

In order to enhance the effective duty cycle of the CTA, we therefore plan to support the short-term scheduling system with *online dynamic scheduling* capabilities. This system should either enable observation of sources visible under good atmospheric conditions over those covered by clouds or aerosol layers, or otherwise select sources according to minimal impact on the later offline data selection, given a respective atmosphere. Such an intelligent system shall be able to judge at each moment whether the observation requirements are still met, or whether a different source with less strict requirements should be observed instead.

Preferences of one part of the sky over another might be issued in the case of cirrus clouds which rarely cover the entire sky. In order to avoid inefficient "hole hopping" between clouds, a minimum prediction of the movement of such locally limited clouds is necessary, given historical knowledge and the wind direction and speed at the altitude of a cloud, e.g. from the GDAS or similar. See [57] for first attempts in this direction.

For such pointing forecasts, scanning instruments and All-Sky-Cameras (ASC) [63] are suitable. Active scanning instruments should employ wavelengths which do not interfere with the CTA cameras, such as *Ceilometers* (operating at 905 nm or 1064 nm [64]). ASCs which have been largely used during the CTA site selection process [65], are becoming more and more standard tools for cloud detection at world-class astronomical observatories [66, 67]. These highly sensitive devices are able to detect even fine cirrus, by comparing detected light fluxes from stars with their catalog values. Exposure times of less than one minute are possible, and provide almost contemporaneous cloud maps. Only cloud height cannot be measured accurately, a task left to the ceilometers.

The scenario is then designed in the following way: the ASC takes a picture of the full sky and converts the result in an atmospheric extinction map (AEM). A cloud recognition program applies a threshold to the AEM and decomposes the result in ellipses, yielding the ASC Clouds List (CL). A global offset (e.g. ground layer enhancement) might be part of the decomposed picture and form the first ellipse extending over the complete FOV of the ASC. If available, the knowledge from available Raman LIDAR shots may be included in the decomposition. The Ceilometer receives the full list of possible observation targets in the short-term scheduler and picks those which are close enough to a cloud in the CL as to get possibly affected by it. After discarding those clouds that remain orphan of associated CTA observation targets, the Reduced ASC Clouds List (RCL) is obtained. The Ceilometer points to the center of each cloud in the RCL and takes an extinction profile[8]. The mean altitude of each cloud is added to the RCL. In the case of more than one layer, only the lower one is picked. A now-cast of wind speed and direction at the relevant cloud altitudes is obtained from GDAS and added to the RCL. Using the RCL and historical knowledge, each cloud in the RCL gets propagated in time for a duration corresponding to next scheduling block, and a probability is calculated for each associated CTA observation target to get covered by the cloud. In case the probability lies above a certain threshold, the expected observation parameters for the next scheduling block: reduction of accessible energy range, degradation of energy and pointing resolution and effective areas are calculated. The list of targets with associated cloud coverage probabilities and performance degradation parameters gets returned to the CTA scheduler for evaluation, before the next scheduling block is executed.

5 Conclusions

While two years ago an atmospheric calibration strategy for the CTA was still being designed [68], we are now constructing and implementing a clear strategy for such a task. Each part of the mentioned ingredients: molecular density profiles, molecular absorption, aerosol extinction, but also systematic differences between MC simula-

[8]This way, it can be always ensured that a Ceilometer shot does not disturb observations from close-by optical telescopes, sensitive to the Ceilometer wavelength.

tion codes [69, 70] and those allowed by simplifications in advantage of simulation speed, is required to contribute by less than 2–3% to the overall uncertainty budget, limited in total to about 7–8%.

With these choices, the CTA will provide much more accurate source spectra (including spectral features), light curves and source morphology measurements than its predecessors and enable the start of the VHE precision gamma-ray astronomy era.

References

[1] Actis, M. et al. *Design concepts for the Cherenkov Telescope Array CTA: an advanced facility for ground-based high-energy gamma-ray astronomy*, Exp. Astron. **32** 193 (2011)

[2] Wakely S. P. & Horan D. *TeVCat: An online catalog for Very High Energy Gamma-Ray Astronomy*, Proc. 30th ICRC **3** 1341 (2008) http://tevcat.uchicago.edu/

[3] Bastieri D. et al. *Energy Calibration of Cherenkov Telescopes using GLAST Data*, Proc. 30th ICRC **3** 1555 (2008)

[4] Meyer M. et al. *The Crab Nebula as a standard candle in very high-energy astrophysics*, A&A **523** A2 (2010), arxiv:1008.4524

[5] Aleksić J. et al. *The major upgrade of the MAGIC telescopes, Part II: A performance study using observations of the Crab Nebula*, Astrop. Phys. **72** 76 (2016), arXiv:1409.5594.

[6] Aharonian F. et al. *Observations of the Crab nebula with HESS*, A&A **457** 899 (2006)

[7] Gaug M. et al. *Calibration strategies for the Cherenkov Telescope Array*, Proc. SPIE Astron. Tel. + Instr. **#9149-45** (2014)

[8] Bernlöhr K. *Impact of atmospheric parameters on the atmospheric Cherenkov technique*, Astrop. Phys. **12** 255 (2000) astro-ph/9908093

[9] Garrido D. et al. *Atmospheric Aerosols at the MAGIC Site*, Proc. 33rd ICRC **0465** (2013) arXiv:1308.0473

[10] Nolan S. J. et al. *Detailed studies of atmospheric calibration in imaging Cherenkov astronomy*, Astrop. Phys. **34** 304 (2010) arXiv:1009.0517

[11] Sobczyńska D.& Bednarek W. *The impact of clouds on image parameters in IACT at very high energies*, Proc. 2nd AtmoHEAD Conf., EPJ **89** 03009 (2015)

[12] Fruck C.& Gaug M. *Atmospheric monitoring in MAGIC and data corrections*, Proc. 2nd AtmoHEAD Conf., EPJ **89** 02003 (2015)

[13] de los Reyes R. *Atmospheric monitoring in H.E.S.S.*, these proceedings

[14] Hahn J. & de los Reyes R. *Atmospheric Monitoring in H.E.S.S.*, Proc. 2nd AtmoHEAD conf., EPJ **89** 02002 (2015)

[15] Hildebrand et al. *FACT - Measuring Atmospheric Conditions with Imaging Air Cherenkov Telescopes*, Proc. 33rd ICRC **0709** (2013)

[16] Dorner D. et al. *A method to correct IACT data for atmospheric absorption due to the Saharan Air Layer*, A&A **493** 721 (2009) arXiv:0808.0279v2

[17] Fruck C. *A novel LIDAR-based Atmospheric Calibration Method for Improving the Data Analysis of MAGIC*, Proc. 33rd ICRC **1054** (2013) arXiv:1403.3591

[18] Font L. *Monitoring and Calibration of the Atmosphere in MAGIC*, Proc. 33rd ICRC **0090** (2013)

[19] Fruck C. *The Galactic Center Resolved with MAGIC and a new technique for Atmospheric Calibration*, PhD Thesis, TU München (2015), https://mediatum.ub.tum.de/node?id=1237928

[20] Acharya B. S. et al. *Seeing the High-Energy Universe with the Cherenkov Telescope Array - The Science Explored with the CTA*, Astrop. Phys. **43** 3 (2013)

[21] Aleksić J. et al. *Optimized dark matter searches in deep observations of Segue 1 with MAGIC*, JCAP **2** 8 (2014) arXiv:1312.1535

[22] Aleksić J. et al. *Measurement of the Crab Nebula spectrum over three decades in energy with the MAGIC telescopes* J. HE Astroph. **5** 30 (2015) arxiv:1406.6892

[23] Abramowski A. et al. *Measurement of the extragalactic background light imprint on the spectra of the brightest blazars observed with H.E.S.S.* A&A **550** A4 (2014) arXiv:1212.3409

[24] García-Gil A. et al. *Atmosphere Extinction at the ORM on La Palma: A 20 yr Statistical Database Gathered at the Carlsberg Meridian Telescope* PASP **122** 1109 (2010)

[25] Stull R. B. *An Introduction to Boundary Layer Meteorology*, Kluwer Acad. Publ. Dordrecht (1988)

[26] Elterman L. *Parameters for Attenuation in the Atmospheric Windows for Fifteen Wavelengths*, Appl. Opt. **6** 745 (1964)

[27] Gaug M. *Site characterization for CTA-N*, these proceedings

[28] Yu H. et al. *The fertilizing role of African dust in the Amazon rainforest: A first multiyear assessment based on data from Cloud-Aerosol Lidar and Infrared Pathfinder Satellite Observations*, Geoph. Res. Lett. **42** (6) 1984 (2015)

[29] Pachauri R. K. & Meyer L. A. (eds.) *IPCC 2014: Climate Change 2014: Synthesis Report. Contribution of Working Groups I, II and III to the Fifth Assessment Report of the Intergovernmental Panel on Climate Change* IPCC Geneva, Switzerland (2014)

[30] Dubovik O. et al. *Application of spheroid models to account for aerosol particle nonsphericity in remote sensing of desert dust*, J. Geoph. Res. **111** D11208 (2006)

[31] Kandler K. et al. *Chemical composition and complex refractive index of Saharan Mineral Dust at Izaña, Tenerife (Spain) derived by electron microscopy*, Atm. Env. **41** (37) 8058 (2007)

[32] Levoni C. et al. *Atmospheric aerosol optical properties: a database of radiative characteristics for different components and classes*, Appl. Opt. **36** 8031 (1997)

[33] Müller D et al. *Aerosol-type-dependent lidar ratios observed with Raman lidar*, J. Geoph. Res. **112** D16202 (2007)

[34] Ansmann A. et al. *Independent measurement of extinction and backscatter profiles in cirrus clouds by using a combined Ramanelastic-backscatter lidar*, Appl. Opt. **31** (33) 7113 (1992)

[35] Andrews E. et al. *Climatology of aerosol radiative properties in the free troposphere* Atm. Res. **102** (4) 365 (2011)

[36] Haffke M. *Atmosphere Is MAGIC. Berechnung und Implementierung neuer Atmospharenmodelle in die MAGIC-Monte-Karlo-Kette*, Dipl. thesis Univ. Dortmund, Germany (2007)

[37] Marín J. et al. *Density profile characterization and modeling at Paranal and Armazones 2k sites*, these proceedings

[38] Vrastil M. *Overview of Atmospheric Simulation Efforts in CTA*, these proceedings

[39] Daniel M. *Application of radiosonde data to VERITAS simulations*, Proc 30th ICRC **3** 1329 (2008)

[40] Lakićević M. et al. *Atmospheric conditions at Cerro Armazones derived from astronomical data*, A&A **588** A32 (2016)

[41] Thompson A. M. et al. *Southern Hemisphere Additional Ozonesondes (SHADOZ) 1998–2004 tropical ozone climatology: 3. Instrumentation, station-to-station variability, and evaluation with simulated flight profiles*, J. Geoph. Res.: Atm. **112** D03304 (2007)

[42] Cuevas E. et al. *Surface O3 characterization in the sub- tropical North Atlantic troposphere*, Atm. Chem. Phys. **13** 1973 (2013)

[43] Haynes P. H. & McIntyre M. E. *On the conservation and impermeability therorems of potential vorticity*, J. Atm. Sci. **47** 2021 (1990)

[44] Rondanelli R. et al. *Rapid changes in ozone mixing ratios at Cerro Tololo (30°10′S, 70°48′W, 2200 m) in connection with cutoff lows and deep troughs*, J. Geoph. Res.: Atm. **107** (D23) ACL 6-1 (2002)

[45] Anet J. G et al. *Surface ozone in the southern hemisphere: 20 years of data from a site with a unique setting in El Tololo, Chile*, Atm. Chem. Phys. **2016** 1 (2016)

[46] Škerlak B. et al. *A global climatology of stratosphere–troposphere exchange using the ERA-Interim data set from 1979 to 2011*, Atm. Chem. Phys. **14** 913 (2014)

[47] Neu J. L. et al. *Tropospheric ozone variations governed by changes in stratospheric circulation*. Nature Geosc. **7** (5) 340 (2014)

[48] Kimeswenger S. et al. *The Innsbruck/ESO sky models and telluric correction tools - The possibility of atmospheric monitoring for Čerenkov telescopes* Proc. 2nd AtmoHEAD conf., EPJ **89** 01001 (2015)

[49] Valore L. et al. *The ARCADE Raman Lidar system for the Cherenkov Telescope Array*, Proc. 34th ICRC (2015) arXiv:1509.02681

[50] Iarlori M. *Experience with Raman LIDARs*, these proceedings

[51] Doro M. et al. *Status and motivation of Raman LIDARs development for the CTA Observatory*, Proc. 1st AtmoHEAD Conf. (2014) arXiv:1402.0638

[52] Eizmendi M. *IFAE-UAB Raman LIDAR Link Budget and Components*, Master Thesis UPC & UAB (2011) http://upcommons.upc.edu/handle/2099.1/12959

[53] Fomin V. P. et al. *New methods of atmospheric Cherenkov imaging for gamma-ray astronomy. I. The false source method*, Astrop. Phys. **2** 137 (1994)

[54] Prouza M. et al. *FRAM The Robotic Telescope for the Monitoring of the Wavelength Dependence of the Extinction: Description of Hardware, Data Analysis, and Results*, Adv. in Astron. **2010** 849382 (2010)

[55] Ebr J. *Aerosol measurements with FRAM telescopes*, these proceedings

[56] Janecek P. *CTA FRAM update*, these proceedings

[57] Adam J. *Allsky Cameras and Cloud Detection*, these proceedings

[58] Sottile G. et al. *UVSiPM: A light detector instrument based on a SiPM sensor working in single photon counting*, Nucl. Phys. B Proc. Supp. **239** 258 (2013) arxiv:1305.2699

[59] Maccarone M. C. *Auxiliary instruments for the absolute calibration of the ASTRI SST-2M prototype for the Cherenkov Telescope Array*, Proc. SPIE Astron. Tel. + Instr. **#9149-44** (2014)

[60] Stepanik S. *Atmospheric monitoring using Cherenkov Transparency Coefficient*, these proceedings

[61] Aleksić J. et al. *Phase-resolved energy spectra of the Crab pulsar in the range of 50-400 GeV measured with the MAGIC telescopes*, A&A **540** A69 (2012) arXiv:1109.6124

[62] Acero F. et al. *Localizing the VHE γ-ray source at the Galactic Centre*, MNRAS **402** 1877 (2010) arXiv:0911.1912

[63] Mandát D. *All Sky Camera for CTA Site characterization*, these proceedings

[64] Münkel C. et al. *Retrieval of mixing height and dust concentration with lidar ceilometer*, Boundary-Layer Meteorol. **124** 117 (2007)

[65] Mandát D. et al. *All Sky Camera instrument for night sky monitoring*, Proc. 1St AtmoHEAD Conf. (2014) arXiv:1402.4762

[66] Nemiroff R. J. et al. *Educational Aspects of the CONCAM Sky Monitoring Project*, AAS Abs. **32** #120.04 (2000)

[67] Smith R. et al. *The Tololo All Sky Camera*, Sci. Detect. f. Astron., Astr. & Space Sc. Lib. **300** 379 (2004)

[68] Doro M. et al. *Strategy implementation for the CTA Atmospheric monitoring program*, Proc. 2nd AtmoHEAD Conf., EPJ **89** 02005 (2014)

[69] Heck D. et al. *CORSIKA: An Air Shower Simulation Program*, Astr. Src Code Libr. **ascl:1202.006** (2012)

[70] Kertzman M. P. & Sembroski G. H. *Computer simulation methods for investigating the detection characteristics of TeV air Cherenkov telescopes*, NIM A **343** 629 (1994)

EPJ Web of Conferences **144**, 01004 (2017) DOI: 10.1051/epjconf/201714401004
AtmoHEAD 2016

Cloud Detection and Prediction with All Sky Cameras

Jan Adam*, *Jens* Buss, *Kai* Brügge, *Max* Nöthe, *Wolfgang* Rhode
[1] *Technische Universität Dortmund*

Abstract. Atmospheric monitoring is a field of special importance for astroparticle physics, especially for Imaging Atmospheric Cherenkov Telescopes (IACTs) as clouds will absorb and scatter the Cherenkov photons of air showers. Conventional tools used for atmospheric monitoring (e.g. LIDAR) are very expensive and monitor only a small part of the sky at once. Therefore, they are not suitable to perform a wide scan of the sky which is necessary to detect clouds in advance.
This article gives a short overview about a method that uses an all sky camera with a 180° field of view to identify the cloud distribution by measuring the absorption of star light. It can be used to assign a sky quality rating to single spots, arbitrary regions or the whole sky at once within a 1 min exposure time. A cloud map can be created from the available data that can be used to determine shape and dimension of clouds and to predict their movement. The resulting data can be used by a scheduling algorithm or the operating crew to point the telescope to a different source before the current source gets covered by clouds. The all sky cameras used so far are located on La Palma at the observatory Roque de los Muchachos close to the telescopes FACT and MAGIC and the planned northern CTA site.

1 Introduction

A precise cloud forecast would be very valuable for the operation of IACTs. If the time was known when a source will cover up, the telescope could switch to another source in advance and continue taking data. The detection of clouds at night using a camera is difficult as they do not emit light on their own and are as dark as the night sky unless they refract moon light and appear as white objects in the camera image. The number of visible stars in the image is therefore a good approximation of the cloud coverage as their appearance does not depend on ambient light or wind conditions.

Regions with visible stars can be marked as not cloudy but a cloud will not always cover stars completely. A partial absorption of light will not be recognized if stars have only two states: 'visible' or 'not visible'. This effect makes it very difficult to reconstruct the smooth shape of a cloud or detect thin clouds in the first place.

2 Image Analysis

Initially, a star catalog is used to find the positions in the image where a star is expected to show. The actual detection of a star is done by convolving the image with a Laplacian of Gaussian filter[3]. This filter is ideal for star detection in long exposure images as it is rotational invariant (anisotropic) and robust against image noise. Figure 1a shows a plot of the filter response R of stars and their magnitude during a starry night. The filter response is correlated to the star magnitude. This allows to use an exponential function:

$$f(x) = 10^{m\cdot x+b} \tag{1}$$

to define an upper and a lower limit for the response of a star (green and red line). The parameters of the green curve were choses such that all stars lie above this curve in a starry night (cf. Fig. 1a).

$$m_u = -0.40 \qquad b_u = 4.45 \tag{2}$$

Stars above this threshold are unaffected by clouds while stars below indicate that a cloud absorbed a portion of their light.

A cloudy night is depicted in Figure 1b and defines the lower threshold for the filter response. The parameter of the red curve were chosen such that the majority of stars lies below this curve in a cloudy night.

$$m_l = 0.0 \qquad b_l = 1.75 \tag{3}$$

Stars below the red curve are not visible at all. Combining both thresholds defines the visibility υ of a star as its relative position between upper and lower threshold in logarithmic space. For example: in Figure 1a the star with magnitude ≈ 0 lies above the green curve, thus its visibility is 1. In Figure 1b its visibility is 0 because it lies below the red curve. If it had a kernel response of 10^4 its position between red and green curve would be $\approx$90 %, thus its visibility would be 0.9. Equation 4 describes this

*e-mail: jan.adam@tu-dortmund.de

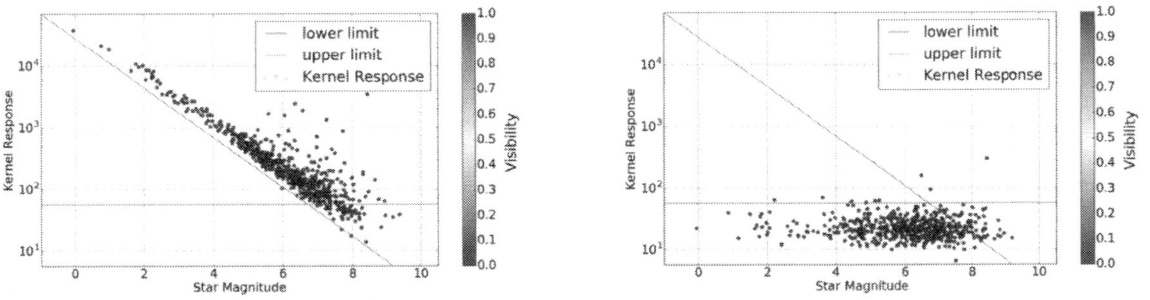

(a) Starry night

(b) Cloudy night

Figure 1: Distribution of the kernel response of stars in a good and a cloudy night. Upper limit is defined such that the majority of stars lie above this line in a starry night and in a cloudy night they should be below the lower limit. Stars with mag > 6.7 are always colored red because once the upper limit crosses the lower limit, the visibility is no longer defined. In the actual analysis, these stars were removed from the catalog and had no influence.

in a mathematical way. The visibility is

$$v = \begin{cases} 1, & \text{for } R > 10^{m_u \cdot \text{mag} + b_u} \\ 0, & \text{for } R < 10^{m_l \cdot \text{mag} + b_l} \\ \frac{R - 10^{m_l \cdot \text{mag} + b_l}}{10^{m_u \cdot \text{mag} + b_u} - 10^{m_l \cdot \text{mag} + b_l}}, & \text{otherwise} \end{cases} , \quad (4)$$

with the slope of the upper threshold m_u and lower one m_l and their y-axis intersection b_u and b_l, respectively, and the star magnitude *mag*.

Averaging the visibility of the stars i in a certain area, makes the algorithm more robust (cf. eq. 5). This area can be chosen arbitrary and may correspond to the telescope's field of view but choosing a wider area that contains more stars yields more robust results. Bright stars are less likely to be affected by a cloud than faint stars therefore a weighting factor[1] of 2.5^{mag} was used to differentiate between thin and thick clouds as faint stars will be covered completely by both.

The averaged star visibility A is defined as:

$$A = \frac{\sum_i^{\text{all stars}} v \cdot 2.5^{\text{mag}_i}}{\sum_i^{\text{all stars}} 2.5^{\text{mag}_i}} \quad (5)$$

3 High Zenith Angle Correction

At high zenith angles the star's filter response decreases and even drops below the upper threshold. Therefore, it appears that the horizon is always covered by clouds. The cause of this phenomenon is most likely either vignetting of the fish-eye lens [4] or absorption and scattering by the atmosphere's natural aerosols. Although, the real cause was not identified yet, the light loss has to be corrected for the algorithm to work reliably.

At high zenith angles the light has to travel a longer way through the atmosphere and thus more photons can be absorbed or scattered. If the absorption is very small, the transmission can be described by the Beer-Lambert law:

$$I(X) = I_0 e^{-kX} \quad (6)$$

with I_0 the initial intensity, $I(X)$ the remaining intensity after distance X and k the absorption coefficient. X is given in units of air mass and by applying the law of cosine on a spherical earth model, the airmass X can be calculated by:

$$X = \sqrt{\left(\frac{R_E + y_{\text{obs}}}{y_{\text{atm}}}\right)^2 \cos^2\theta + \frac{2R_E}{y_{\text{atm}}^2}(y_{\text{atm}} - y_{\text{obs}}) - \left(\frac{y_{\text{obs}}}{y_{\text{atm}}}\right)^2 + 1} - \frac{R_E + y_{\text{obs}}}{y_{\text{atm}}} \cos\theta. \quad (7)$$

with observer altitude $y_{\text{obs}} = 2.2\,\text{km}$, earth radius $R_E = 6378\,\text{km}$ and atmosphere height $y_{\text{atm}} = 9.5\,\text{km}$.

To estimate the atmospheric absorption coefficient k, images of three consecutive nights with good weather were used. The images were taken at an interval of 15 min. These measurements of stars at different zenith angles allow to observe the decreasing filter response. A fit of equation 6 to a star's data points yields a k for every single star. The absorption coefficient should be equal for all stars but is spread with $k = 0.35 \pm 0.27$. No correlation to the star's magnitude was observed. The 75 % quantile of k was used to correct the response of all stars. This leads to an over correction of $\approx\frac{3}{4}$ of all stars but a cloud will still block enough light for a notable effect. Under corrected stars, however, induce fake clouds all the time. Therefore, an over correction is the better choice if one corrects all stars with a constant k. The correction of the kernel response was performed before calculating the star visibility.

4 Cloud Map

The star visibility is only defined at the location of a star. To visualize clouds, it is necessary to develop a parameter that is homogeneous in space. One could achieve this by

[1] Brightness factor between consecutive magnitudes $\sqrt[5]{100} \approx 2.5$.

DOI: 10.1051/epjconf/201714401004

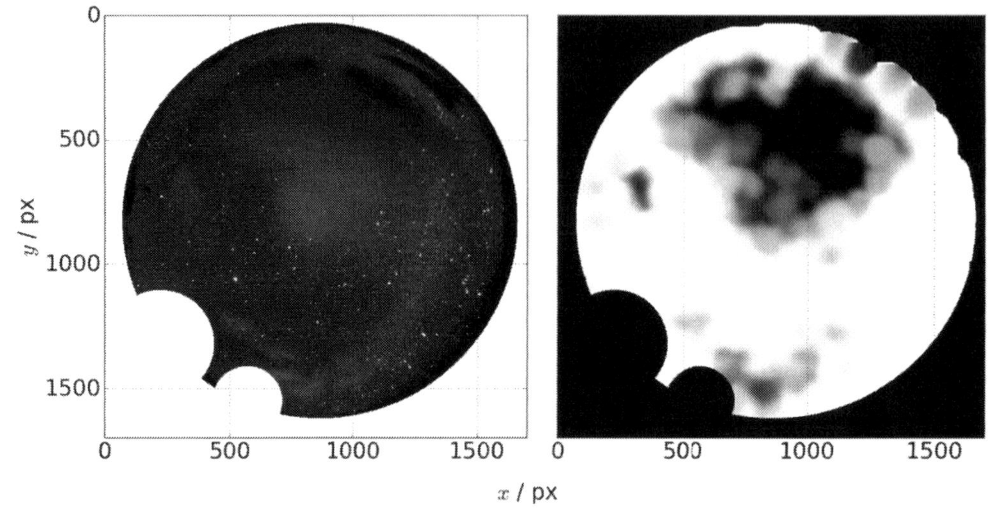

Figure 2: All sky camera image (left) and the corresponding cloud map (right) that results from the algorithm explained in section 4 and Figure 3. Dark pixels show cloudy regions. The shape of the cloud was reconstructed very detailed and even thin cloud areas can be identified. Clouds close to the horizon were not reconstructed correctly because the number of available stars was too low. In the lower left, the MAGIC telescopes were removed from the image.

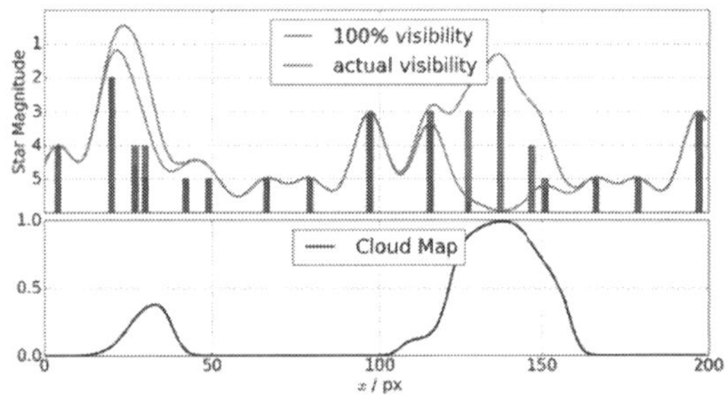

Figure 3: Explanation of the Cloud map algorithm. The upper plot is a slice of the 2D histogram discussed in the text. The bin size correlates to the star magnitude and the level of the green color displays the fraction of their visibility. The curves are the result of Gaussian smoothing. The normalized difference between both curves is the cloud map displayed in the lower plot. Here, two clouds are visible as peaks with different intensities.

calculating the averaged star visibility for every pixel of the image within some radius r but this is computationally very expensive.

A less expensive algorithm that also uses a filter kernel is visualized in Figure 3. Every star was filled into a 2D histogram based on its x and y position in the image. Each entry then was weighted by the star's visibility and the magnitude factor 2.5^{mag} (green bins). Then, the histogram was treated as a regular image and a Gaussian filter was applied to smear the discrete bin values (green curve). However, this 'map' is biased to the density of stars. To normalize this curve, another 2D histogram was created with $v = 1$ for all stars (red bins and curve). After applying the Gaussian kernel, the normalized difference between both curves can be calculated (black curve in lower plot). This is the cloud map used in all further steps. It visualizes the cloud coverage and differentiates between thick and thin clouds but is not biased by changing light conditions like the original camera image.

This implementation takes only seconds to compute a cloud map for the whole sky whereas a LIDAR measurement is in the order of minutes for a spot-measurement. This map can then be made available for the telescope's operating crew once a minute each night. An example is displayed in Figure 2. The cloud map visualizes the structure of clouds with a very good resolution and even very thin clouds that only block a portion of the star light are visible.

5 Cloud Tracking and Prediction

Identifying single clouds in the cloud map could be done by using a clustering algorithm. Then their direction could be reconstructed by tracking each cloud separately in consecutive images. Clouds, however, can not only change their shape but also split and merge if they move at different velocities.

As a proof of concept, the movement of all clouds was estimated as a unified object instead of identifying and tracking single clouds. The previous and the current

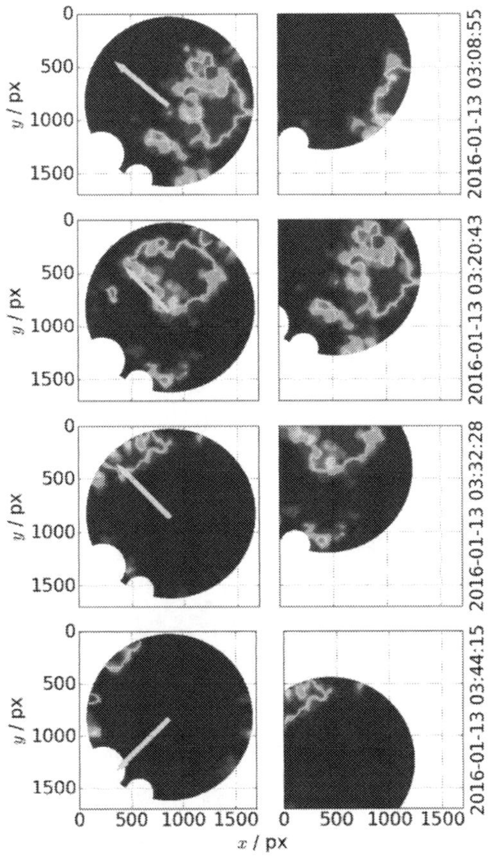

Figure 4: Cloud tracking and prediction. The left column shows cloud maps containing a cloud that moved within approx. 40 min from the lower right of the image to the upper left. The yellow arrow shows the translation that yielded the smallest Euclidean distance between this map and the previous one (wind speed and direction). The right column shows how the previous cloud map was translated to get the best matching position. The tracking fails in the final row because the cloud left the image and no good match was achieved.

cloud map were overlapped and interpreted as n dimensional vectors. The Euclidean distance

$$d = \|u - v\|_2 \tag{8}$$

was calculated between both maps u and v to rate their similarity. By shifting the previous map in x and y direction, the best translation was found by minimizing the Euclidean distance. This translation vector should correspond to the current wind speed and direction and can be used to predict the next position of the clouds if the wind is constant to some degree. Wind speed and direction can also be obtained from wind profiles[5] and could be used to verify the results. The process is displayed in Figure 4 for a cloud that passes through the image.

The algorithm manages to extract the direction of the cloud in the first three rows (yellow arrow). The reconstruction fails if the cloud enters or leaves the image (row four) because then no valid match can be achieved between both maps.

6 Conclusion

Cloud tracking and prediction are promising but need some improvements: Changes of the wind direction could be limited to a few degree within a short period of time such that rapid changes like in row four of Fig. 4 do not occur anymore.

Projecting the cloud map onto the model of the spherical atmosphere will increase the accuracy of the prediction because right now, clouds move faster in the image center due to the fish eye lens.

The usage of wind profiles might be sufficient to calculate wind speed and direction with a good precision such that the cloud map will only be used for detecting clouds and determining their shapes. This also increases the forecast time because calculating the speed of a cloud from the cloud map requires some time to accumulate enough images. By that time the cloud already moved further into the image center and by then it might be too late to switch the telescope to a different source.

The cloud maps provide a detailed visualization of the sky's cloud coverage. In combination with a LIDAR for altitude measurements they can be a valuable tool for future IACTs.

The code used for this analysis was developed at:
https://github.com/tudo-astroparticlephysics/starry_night

Acknowledgement: This work has been supported by the DFG, Collaborative Research Center SFB876, project C3 (http://sfb876.tu-dortmund.de/)

References

[1] M. Will for the MAGIC collaboration, *Atmospheric monitoring at the Site of the MAGIC Telescopes* (2016), these proceedings

[2] Dusan Mandat, *All Sky Camera for the CTA Atmospheric Calibration work package* (2014)

[3] Tinku Acharya and Ajoy K. Ray, *Image Processing, Principles and Applications*, (John Wiley & Sons, Inc., 2005, 452)

[4] Coralie Cauwerts, Magali Bodart PhD, and Arnaud Deneyer, *Comparison of the Vignetting Effects of Two Identical Fisheye Lenses*, (Leukos Volume 8, 2012 - Issue 3, p. 181-203)

[5] https://www.ready.noaa.gov/gdas1.php, GDAS wind profiles

[6] Jan Adam, *Atmospheric Monitoring and Cloud Prediction with All Sky Cameras*, Master's thesis, TU Dortmund (2016)

All Sky Camera for CTA Site characterization

Dusan Mandat[1,*] and *Miroslav* Pech[2]

[1] *Institute of Physics of the Academy of Sciences of the Czech Republic, Na Slovance 2, Prague, Czech Republic*
[2] *RCPTM, Joint Laboratory of Optics of Palacky University and Institute of Physics of CAS, Palacky University, 17. listopadu 12, 771 46 Olomouc, Czech Republic*

Abstract. The All Sky Camera (ASC) for CTA Site characterization is an automated optical instrument providing optical observation of the night sky conditions above the future Cherenkov Telescope Array observatory. It provides the basic sky condition analysis like cloud coverage, cloud position, nightsky background in visible light wave range and basic raw photometry. The result will serve for further Monte Carlo design studies and calculations. The ASC and additional auxiliary instruments (e.g. Ceilometer) will provide input data for smart scheduler during the standard measurement procedure of the CTA observatory.

1 Introduction

The Cherenkov Telescope Array (CTA) [1] will be the next generation of gamma-ray astronomy ground-based observatory for very-high energies. With more than 100 telescopes located in the northern and southern hemispheres, CTA will become the world's largest and most sensitive high-energy gamma-ray observatory. The atmosphere has to be monitored during the observation as the measurement process is strongly sensitive to the current atmospheric condition. There will be a complex program for atmospheric monitoring running within the observatory to study real-time atmospheric parameters to allow eventual changes of the measurement schedules or for correcting the measured data and analysis. The ASC is an instrument for the site characterization during the pre-production and deployment phase and serve for the dynamic scheduler of the CTA observatory during the standard observation.

2 Analysis of clouds above the observatory

The main task of the instrument is the analysis of the clouds on the sky over the CTA observatories before and during the measurement period. The goal is to monitor the clouds condition, changes and a local behavior of the atmosphere in term of cloudiness during the nights. The CTA observatory selected two potential sites for CTA north and south observatory. Both sites were very well investigated and described in the meaning of cloud coverage during the CTA site search campaign. The comparison of the CTA candidate sites was carried out during the period of 2011 - 2015 [2]. The data for the campaign could be used e.g. for the estimation of the cloud-free nights. The cloud-free night could be defined as a night with the cloudiness of the sky (or cloud fraction cf - fraction of the sky covered with the clouds) less than 20% during the whole night. The limit of the cloud-free sky - less 20% of the sky was set as the CTA clear sky criteria during the site search campaign. The comparison is shown in the figure 1. Vertical axis represent the cloudy/clear nights, where 0 means completely covered sky during the whole night, and 1 means completely clear sky during the whole night with respect to the CTA definition. The whole data set of both sites represents 457(red color)/439(blue color) analyzed nights. SITE_1 (red color) shows better sky condition in terms of clear sky comparing to the second one. We do not present the location of the site, it is just a comparison of both future CTA observatories and the cloud condition of the particular site is one of the criteria of the quality of CTA sites. The CTA criteria could be also changed. The figure 2 shows the number of good quality nights with respect to the cloud fraction criteria cf. If assumed the scenario with the cloud fraction e.g. 50 (50% of the sky is covered with the clouds) the total number of good quality nights could significantly increase with respect to the CTA cf criteria (see fig. 2). The result depicts the expected percentage of nights without observation (number of fully covered nights) and nights without any observation losses (completely clear nights).

3 Site characterization

3.1 Clouds cover

The CTA north and south sites were chosen to host new CTA observatory. The average cloud coverage was measured and analyzed [3, 4] during the CTA site search campaign at both locations. The next step of the site characterization is a deep study of the night sky condition with respect to seasonal variations, local anomaly, and standard

*e-mail: mandat@fzu.cz

DOI: 10.1051/epjconf/201714401005

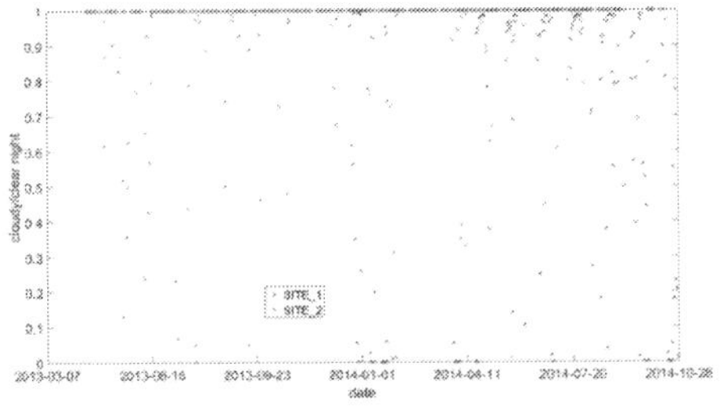

Figure 1. The comparison of cloudless night of CTA sites. 0/1 - fully covered/clear sky during the whole night.

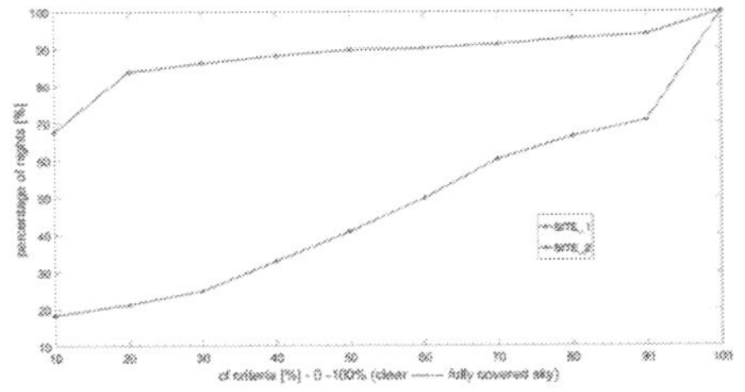

Figure 2. The fraction of fully clear/covered nights as a function of cf (cloud fraction) criteria of CTA sites.

nature of the clouds above the observatory. We installed a new generation of ASC in 10/2015 at CTA north and 11/2015 at CTA south site. The new generation of the camera contains additional CCD cooling system and photometric Johnsons filters comparing to the first ASC generation dedicated for the site search campaign. The resolution of the CCD chip of the camera (Moravian Instruments G2-4000) is 2056 × 2062 pixels, the angular resolution is 0, 1° and the lens f number is 2.8 (Sigma 4.5/2.8 EX DC). We observe stars up to 7 star magnitude in zenith with 30 sec. exposure time typically. The figure 3 shows the result of cloud analysis. The green/red circles show detected/not-detected (covered by cloud) stars. The figure represents three subsequent images. The clouds are visible and the red arrows show the estimated cloud movement direction. Further analysis will focus on new method that should consist in estimating the cloud movement direction using a set of previous analyzed clouds maps. The method should be able to determine the location and the time of clouds and predict if the target source will be observable or not, providing input to the scheduler.

3.2 Star photometry

The new generation of the ASC contains 5 position filter wheel for 1.25" threaded cells with BVR Johnsons filter installed. The Johnsons BVR filter allows us to measure the flux from the known stars and compare the signal with the catalog for different wave range. TYCHO2 star catalog is used (for the astrometry and clouds analysis) for the star photometry. The comparison of the flux measured

Figure 3. Cloud analysis of three subsequent images during one cloudy night. The green/red circles represent detected/not-detected stars. The clouds are also visible due to the moon in the field of view (the moon light is scattered from the clouds). The clouds are moving top-down direction (red arrows). The direction was estimated by observer.

by ASC and TYCHO2 value could be used for the RAW estimation of the integral atmospheric attenuation in the direction to the observed star as shown in figure 4. The red dashed curve represents the theoretical extinction of the source for different zenithal angle. It is the zenithal atmospheric brightness - extinction at zenithal angle. The dots with different colors (one color per one star at different zenithal angle) depict the differences between the measured star flux (using ASC) and the catalog value of individual analyzed stars. The cloud-free nights were selected for this analysis only. The offset of the theoretical (catalog) and measured magnitudes is caused by the attenuation of the ASC optics (namely vignetting) and its non-homogeneous spectral transmission of the ASC lens across the FoV. Higher is the zenith angle, greater is the difference (figure 4). To mitigate the offset caused by the optical elements a local catalog will be generated for particular site, especially during the long term site characterization campaign. Thus the final analysis for routine observation will be independent of a star catalog and spectral correction.

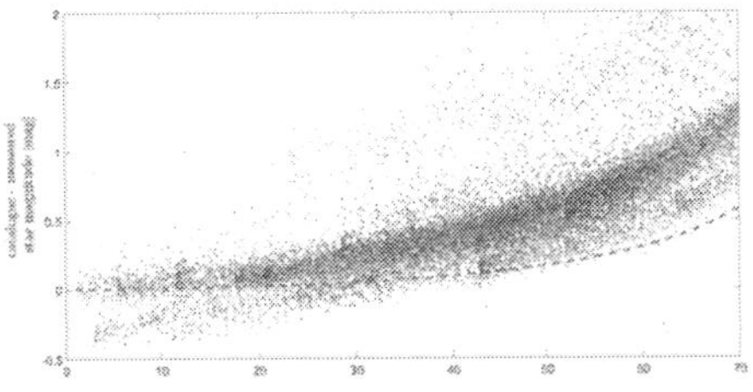

Figure 4. The result from a set of 300 clear sky images. The dashed red curve represents a zenithal atmospheric brightness. The color points represent the differences between signal measured (in magnitudes) by the ASC and the star catalog value for different zenithal angles (one star - one color).

Acknowledgments

We gratefully acknowledge support from the agencies and organizations listed under Funding Agencies at this website: http://www.cta-observatory.org/. The authors are grateful for the support by the grants of the Ministry of Education, Youth and Sports of the Czech Republic, projects LE13012, LG14019 and LM2015046.

References

[1] B.S. Acharya, et. al., Introducing the CTA concept, Astroparticle Physics **43**, 3–18 (2013).

[2] D. Mandat, for the CTA Consortium, All Sky Cameras for the characterization of the Cherenkov Telescope Array candidate sites, in Proceedings of the 33rd International Cosmic Ray Conference, (2016).

[3] R. D. Piacentini, et. al., Selection of astrophysical/astronomical/solar sites at the Argentina East Andes range taking into account atmospheric components, Advances in Space Research, **57** (12), 2559-2574 (2016).

[4] G. Tovmassian, Selection of astrophysical/astronomical/solar sites at the Argentina East Andes range taking into account atmospheric components, Publications of the Astronomical Society of the Pacific, **128** (961), 035004 (2016).

[5] G. Martucci, C. Milroy, C. D. O'Dowd, Detection of Cloud-Base Height Using Jenoptik CHM15K and Vaisala CL31 Ceilometers, **21** (2), 305 - 318 (2010).

[6] P. Janecek, M. Prouza, J. Ebr, FRAM for CTA, EPJ Web of Conferences, **89** (2), 1 - 4 (2015).

DOI: 10.1051/epjconf/201714401006

Atmospheric aerosol measurements from satellites

Nataliia Borodai[1,*]

[1] *Institute of Nuclear Physics Polish Academy of Sciences, PL-31342 Krakow, Poland*

Abstract. Satellite instruments are commonly used for measurements of aerosol optical properties in the atmosphere. In this work we use such measurements for ultra-high energy cosmic ray studies. We introduce a method of preparation of aerosol maps for the location of the Pierre Auger Observatory. For these maps the data from the Ozone Monitoring Instrument (OMI) of the Aura satellite were analyzed for the period 2004-2016. Such aerosol properties as the aerosol optical depth (AOD) and aerosol types were retrieved. The seasonal dependence of aerosol distribution is also studied.

1 Introduction

Aerosols affect the radiation budget of the Earth's atmosphere directly by scattering and absorbing solar radiation. They also influence indirectly clouds, modifying their optical properties and lifetime. Aerosols in the atmosphere influence the detection of extensive air showers developing in the atmosphere. This is why it is important for the cosmic ray physics to know the distribution of aerosols in the atmosphere.

Currently there are several instruments for aerosol measurements used at the Pierre Auger Observatory [1],[2]. Among them are Central Laser Facility (CLF) [3] and eXtreme Laser Facility (XLF), LIDAR stations located near each fluorescence detector building [4], and others. However, these instruments do not give the global picture of aerosols over the Observatory. They measure aerosols only along a certain direction. Measurements done using satellite data allow one to produce aerosol maps, which show the periodic increases and decreases of aerosols in time over the whole Observatory. This is especially important since the fluorescence detectors of the Observatory view the sky over large areas and the aerosol concentrations may vary at different locations within their field of view.

There are a number of satellites with instruments capable of measuring aerosols in the Earth's atmosphere. Among these instruments is the Ozone Monitoring Instrument (OMI) [5] aboard NASA's Earth Observing System's (EOS) Aura satellite. Aura was launched on July 15, 2004, and OMI has collected data since August 9, 2004.

The Aura spacecraft is in a polar sun-synchronous orbit at an altitude of 705 km and has a daily global coverage with on-average 14 orbits. Aura flies in a formation of satellites, called "A-Train" (the Afternoon constellation) [6]. The primary goal of the constellation is to provide observations of clouds and aerosols which are not possible from ground-based instruments.

2 OMI aerosol measurements

The OMI is a nadir-viewing spectrometer. It is a kind of passive imager which measures the amount of solar radiation backscattered by the Earth's atmosphere and surface.

In this study we used reflected solar radiation measurements from the OMI instrument to estimate the aerosol optical depth (AOD) over the Pierre Auger Observatory. The AOD is the aerosol extinction coefficient integrated vertically from the surface to the top of the atmosphere.

2.1 OMI instrument

The OMI instrument measures the solar backscattered radiance in the wavelength range between 270 and 500 nm, with a spectral resolution between 0.42 nm and 0.63 nm. In this wavelength range, the scattering and absorption by atmospheric aerosol particles varies smoothly with wavelength. The wavelength dependence is used to obtain information on the aerosol type, concentration and the size distribution. The nadir pointing telescope of OMI has a very large field of view of 114°, which is used for swath registration, perpendicular to the flight direction of the satellite. Due to so large a field of view, the OMI swath width is 2600 km, which provides daily global coverage. A nominal ground footprint or ground pixel size is 13x24 km^2 at nadir.

Figure 1 represents the real OMI measurements done over the location of the Pierre Auger Observatory on June 9, 2013. Green points represent ground pixel centers, numbers show the sequence of measurements (every 2 sec).

The time of the measurements over the Pierre Auger Observatory is usually the evening time: around 17.00-19.00 UTC (14.00-16.00 local time).

*e-mail: nataliia.borodai@ifj.edu.pl

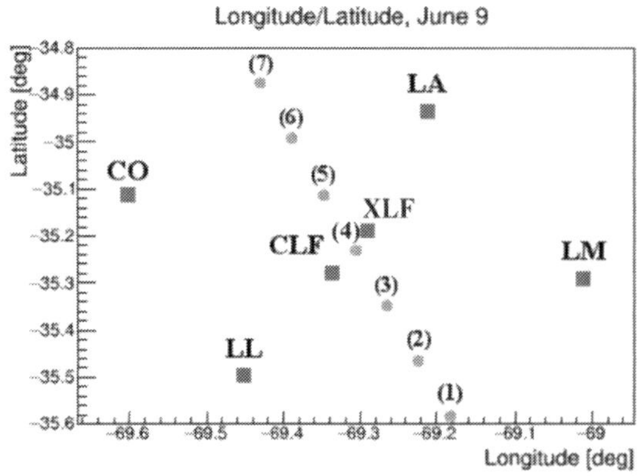

Figure 1. The OMI measurements over the Pierre Auger Observatory performed on June 9, 2013. Red squares correspond to locations of the fluorescence detectors (LA, LM, LL, CO) [2] and the laser facilities (CLF and XLF) at the Pierre Auger Observatory. Green points represent ground pixel centers, the numbers show the sequence of measurements.

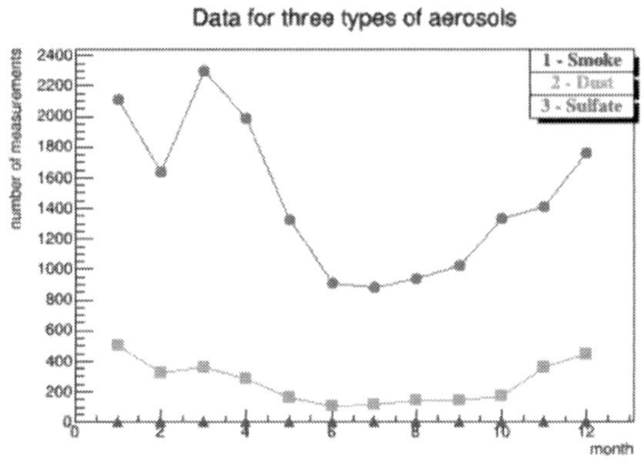

Figure 2. Number of OMI measurements over the Pierre Auger Observatory over years 2004-2016 estimated for three types of aerosols. Blue triangles represent smoke, green squares are dust and red dots are sulfate-based aerosols.

Among OMI capabilities are a measurement of aerosol optical depth (AOD), the ability to distinguish between aerosol types, such as smoke, dust and sulfates, the ability to measure aerosol absorption capacity in terms of aerosol absorption optical depth or single scattering albedo. The AOD and aerosol type determinations are the main properties studied in this work to be used in cosmic ray shower investigations.

2.2 Retrieval algorithms and aerosol types

There are two algorithms [7] used for aerosol retrieval: Near UV aerosol retrieval algorithm (OMAERUV) and Multi-wavelength Algorithm (OMAERO).

OMAERUV is used to retrieve the AOD values over land. It uses as input OMI measured reflectances at 354 and 388 nm to retrieve column atmosphere values of AOD, using a set of 21 aerosol models. The algorithm differentiates the types of aerosols: desert dust, biomass burning carbonaceous particles, sulfate-based aerosols.

OMAERO is used to retrieve the AOD values preferably over ocean. It uses the measurements from 15 wavelengths in the range from 342.5 to 483.5 nm.

In the following we use only the data retrieved using the OMAERUV algorithm as it is more common for over land measurements.

There are three aerosol types distinguished in OMI data in the OMAERUV algorithm: smoke (biomass burning carbonaceous particles), desert dust and sulfate-based aerosols. Taking all the available data for the location of the Pierre Auger Observatory for the years 2004-2016 and applying some cuts (see cuts 1-3 described in section 2.3) Figure 2 was produced. This figure shows the number of satellite measurements of aerosols classified according to their types defined from the OMAERUV algorithm. The number of measurements is plotted vs month to see the seasonal changes of the quantities of aerosols of different types. It appears that sulfate-based aerosols constitute the main part of the aerosols (more than 85%).

According to OMI models the distribution of the sulfate-based aerosols decreases exponentially with height. Other aerosol types have distributions with maximum at some altitude. So, in the following only sulfate-based aerosols are taken into account for AOD estimation.

2.3 Dataset

The OMAERUV algorithm data with wavelength 354 nm are chosen for the following analysis. This wavelength is very close to the one used by the CLF instrument at the Pierre Auger Observatory (355 nm).

There are some standard satellite data quality flags [8], such as: row anomaly correction flag, pixel quality flags, measurement quality flags, algorithm flag. These quality selection criteria should be applied to obtain the OMI level 2 data, which are used in the current analysis.

In summary, to obtain a more clean dataset the following cuts on data are applied: 1. OMAERUV algorithm AOD values are chosen; 2. the data with the wavelength 354 nm are taken; 3. satellite data quality flags are applied; 4. only sulfate-based aerosols are taken for AOD estimation.

In this work the satellite dataset is limited to the temporally averaged data. So, after applying the cuts all available OMI measurements were taken for the period from August 2004 to June 2016 for the region of the Pierre Auger Observatory. Each available data point corresponds to the ground pixel with its own longitude and latitude coordinates (same as shown in Figure 1). All these data points were then mapped according to their positions. The map with the points was then gridded at 0.2°x0.2° to build the aerosol map. Finally, the map was plotted as a 2D histogram smoothed with cell size of about 0.04°x0.04° as shown in Figure 3. The mean <AOD> is provided along with its RMS, standard deviation of the mean and the number of the data points.

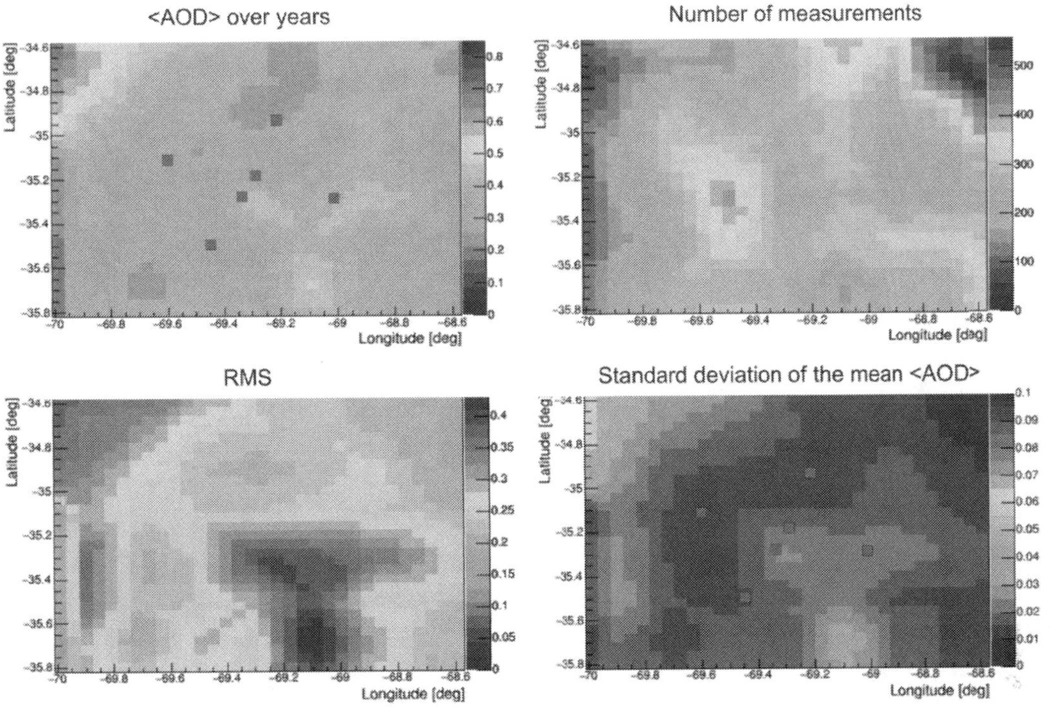

Figure 3. Estimation of <AOD> and its errors for years 2004-2016 for the region of the Pierre Auger Observatory. Red squares represent the fluorescence detectors and the lasers (as shown in Figure 1). Top-left image: average <AOD> map; top-right image: number of AOD measurements averaged over years; bottom-left image: estimated RMS for <AOD>; bottom-right image: estimated standard deviation of the mean <AOD>.

3 Results and discussion

According to the procedure described in section 2.3 the aerosol maps over the Observatory were produced. Average values of <AOD> (top-left image in Figure 3) may suggest that aerosols are inhomogeneous over the Pierre Auger Observatory.

It should be mentioned that the OMI measurements of AOD are not confirmed by the aerosol measurements performed at the Pierre Auger Observatory using CLF (viewed from each of the FD sites), which point out that the aerosols are close to being homogeneous [1]. In addition, <AOD> values obtained from satellite are up to several times larger than the values measured at the Pierre Auger Observatory, and so are inconsistent with the ground-based measurements.

One should keep in mind that OMI data analysis relies heavily on models to describe the state of the atmosphere. A valid question is whether the global atmospheric models used in satellite data reduction describe well the local conditions at the site? It is well known that to correctly describe the real atmosphere it is necessary to account for relatively quick time variations of the state of the atmosphere [9].

A possible explanation for the large absolute values of satellite AOD values can be sub-pixel cloud contamination which was reported by OMI group [8]. It happens because of the relatively large pixel size of the OMI observations. So, the mean AOD can be significantly overestimated. On the other hand, according to [8], the experience with the OMI predecessor satellite spectrometer TOMS (Total Ozone Mapping Spectrometer) suggests that monthly mean <AOD> values do reliably capture variation in the AOD with time.

One more source of uncertainty is the OMI cloud screening scheme, due to which the data with cloud days can be coincidentally included to the analysis. So, in the future analysis the OMI data should be used together with cloud information from other sources to exclude partly clouded OMI scenes that are not recognized as being cloudy by OMI.

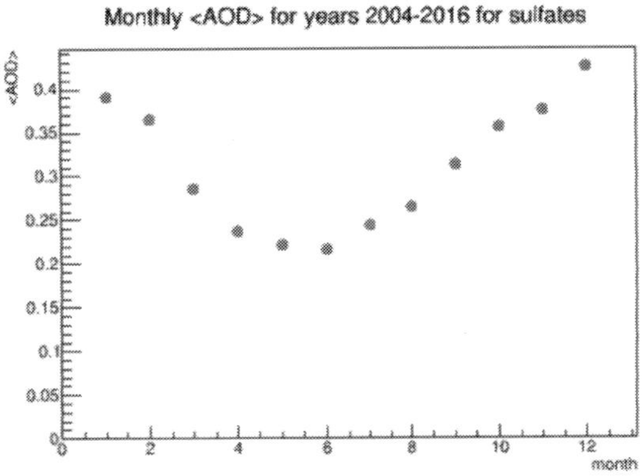

Figure 4. Monthly <AOD> for sulfates averaged for the years 2004-2016.

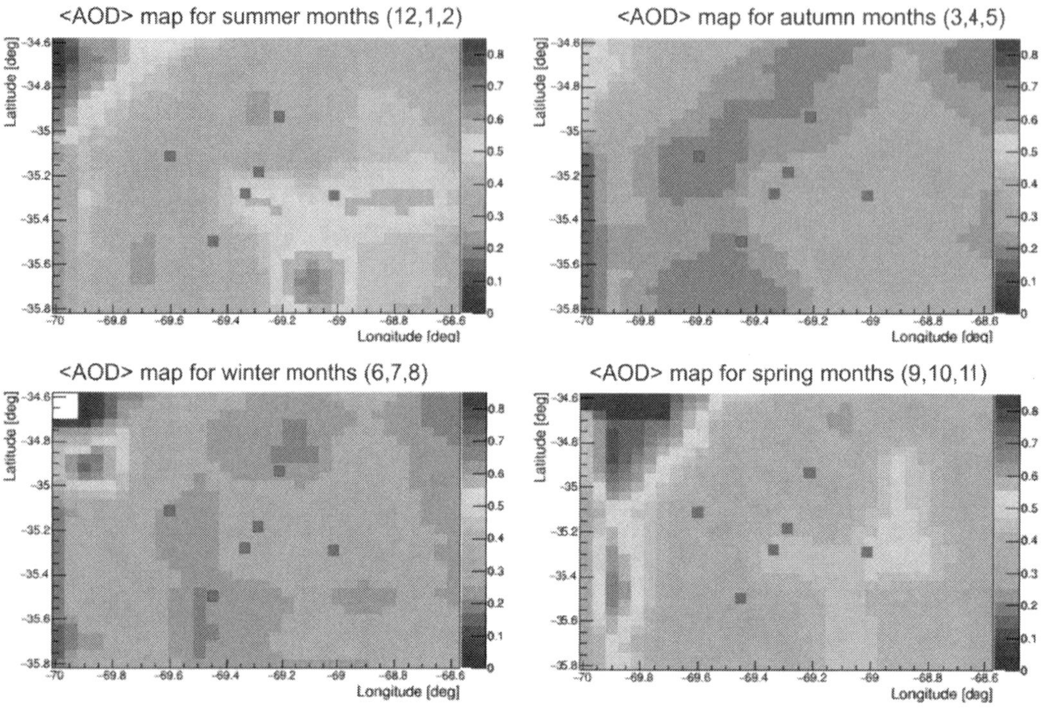

Figure 5. Seasonal <AOD> maps for years 2004-2016 produced for the location of the Pierre Auger Observatory.

However, although the average values of AOD from OMI seem to have problems, it may be interesting to look at seasonal variation of aerosols.

Using the described dataset the monthly average <AOD> was obtained for sulfates (see Figure 4).

In Figure 5 the seasonal changes of the aerosols are presented. These maps correspond to the results presented in Figure 4, showing the highest values of <AOD> during summer and the lowest during winter.

4 Conclusion

This work is an attempt to analyse aerosol data from polar satellites for the purposes of cosmic ray physics. The information obtained from this analysis may cause many questions. Although the OMI satellite-based aerosol measurements point out the inhomogeneity of aerosols over the Pierre Auger Observatory, the question about the existence of this inhomogeneity is still open as the quality of the OMI aerosol data is not sufficient to prove the obtained results. Among the questions there is also one about sulfates: why do we see such a large fraction of sulfate-based aerosol emissions in that region? Can they be due to a large content of volcanic ash in the soil? For the time being, we do not have answers to those questions.

So, further studies of <AOD> are definitely needed. It may be useful to use the cloud and aerosol information from other satellites. Particularly, satellites that fly in the A-Train formation together with Aura satellite possess the instruments for cloud and aerosol analysis, namely: the global imager MODIS on EOS-Aqua, the cloud radar on CLOUDSAT, the aerosol/cloud lidar on CALIPSO.

Acknowledgments

This work was partially supported by the National Science Centre under grant 2013/08/M/ST9/00728.

The author also thanks NASA's Goddard Earth Sciences Data Information Services Center (GES DISC) and the OMI aerosol research team for providing the satellite data used in this study.

References

[1] The Pierre Auger Collaboration, Astropart. Phys. **33**, 108 (2010)

[2] The Pierre Auger Collaboration, Nucl. Instrum. Meth. A **798**, 172 (2015)

[3] B. Fick et al., JINST **1**, P11003 (2006)

[4] S. BenZvi et al., Nucl. Instrum. Meth. A **574**, 171 (2007)

[5] OMI Algorithm Theoretical Basis Document, Volume 1 (ATBD-OMI-01), (2002)

[6] NASA Facts, A-Train Factsheet (Fs-2003-1-053-gsfc), Formation Flying: The Afternoon "A-Train" Satellite Constellation (2003)

[7] OMI Algorithm Theoretical Basis Document, Volume 3 (ATBD-OMI-03), (2002)

[8] Ozone Monitoring Instrument (OMI) Data User's Guide (OMI-DUG-5.0), (2012)

[9] B.Wilczyńska et al., Astropart. Phys. **25**, 106 (2006)

Density profile characterization and modeling at Paranal and Armazones 2k sites

Julio C. Marín[1,2,⋆], *Omar* Cuevas[3,2,⋆⋆], and *Diana* Pozo[1,2,⋆⋆⋆]

[1]*Departamento de Meteorología, Av. Gran Bretaña 644, Playa Ancha, Valparaíso, Chile*
[2]*Centro de Estudios Atmosféricos y Astroestadística, Fac. Ciencias, Universidad de Valparaíso, Chile*
[3]*Instituto de Física y Astronomía, Av. Gran Bretaña 1111, Playa Ancha, Valparaíso, Chile*

Abstract. The Cherenkov Telescope Array (CTA) in the southern hemisphere will be installed at Armazones 2k site in northern Chile. Scarce atmospheric observations are available in the region, particularly radiosonde data. This study analyzes radiosondes launched at Paranal observatory, located at about 21 km from the CTA site, from 24 October and 4 November 2011, to understand the behavior of density in the atmosphere near the CTA site. High-resolution numerical simulations with the Weather Research and Forecasting (WRF) model are validated with Paranal radiosondes to quantify its ability to represent the atmospheric conditions in the region. In addition, the seasonal and diurnal evolution of atmospheric density at the CTA site were studied during 2011 using the high-resolution weather forecasts from the WRF model.

1 Introduction

A number of ground-based imaging atmospheric Cherenkov telescopes (IACTs) have been installed (H.E.S.S. [1]), VERITAS [2]) and MAGIC [3]) to observe gamma rays from galactic and extragalactic sources. The Cherenkov Telescope Array ((CTA) [4]) project is a large array of next-generation very high energy gamma ray telescopes that will improve the sensitivity of the gamma ray domain by installing two arrays in each hemisphere to cover the full sky. The one in the southern hemisphere will be located at Armazones 2k site in northern Chile (Fig. 1).

The atmosphere is used as a calorimeter in ground-based Cherenkov telescopes. Atmospheric density profiles affect the detection of air showers and the transmission of Cherenkov light through the atmosphere. For example, [5] found a 60% difference in Cherenkov light yield between tropical and polar atmospheric profiles, and 15-20% differences at a mid-latitude site due to seasonal variations in atmospheric profiles. Measurements from a radiosonde campaign will provide the information to reduce the uncertainties in the models used to simulate the Cherenkov light at the CTA site. In addition, radiosondes can be used to validate global or mesoscale models, which have been used to characterize sites hosting experiments to detect high energy cosmic rays ([6], [7]).

The arid northern Chile is a region with scarce atmospheric observations. This work aims to provide some information on the atmospheric characteristics at and near the southern CTA site. In order to do that: 1) we will analyze density profiles from radiosondes launched at a nearby site to CTA (Paranal Observatory), 2) validate high-resolution numerical weather simulations with Paranal radiosondes, and 3) study the diurnal and seasonal evolution of the atmospheric density at Armazones 2k site using high-resolution weather forecasts.

⋆e-mail: julio.marin@meteo.uv.cl
⋆⋆e-mail: omar.cuevas@uv.cl
⋆⋆⋆e-mail: diana.pozo@meteo.uv.cl

2 Data and methodology

2.1 Radiosonde campaign at Paranal observatory

The Paranal observatory is located at about 21 km from Armazones 2k site, in the North of Chile. Scarce meteorological observations make difficult to study the region. A radiosondes campaign was conducted from 24 October to 4 November 2011 at Paranal observatory, where 21 RS92-SGP radiosondes were launched at 00 and 12 UTC. In this study, 19 radiosondes were used: 8 launched at 00 UTC and 11 launched at 12 UTC.

2.2 WRF model configuration

The WRF model is a mesoscale numerical weather prediction system that is used both for operational forecasting and atmospheric research. Version 3 of the model [8] was used in this study. The WRF model was run fully compressible, non-hydrostatic, with four nested domains (Fig. 1) and 69 vertical levels. This study evaluates the domain with higher horizontal resolution (d04 at 1 km) centered on the Paranal observatory site. Global Forecast System (GFS) forecasts provided the initial and boundary conditions for WRF simulations every 3 hours. Simulations

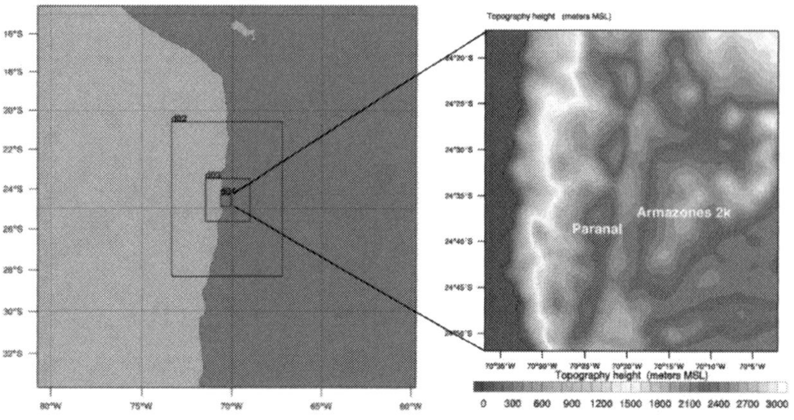

Figure 1. Left: The four nested domains used in WRF simulations in the study. Right panel: zoomed-view of the innermost domain (d4) showing the locations of Paranal and Armazones 2k sites.

were run every day for 24 hours, completing 315 days of 2011. Domain d04 at 1 km horizontal resolution was used in the analysis presented below.

3 Results

3.1 Radiosondes analysis

Figure 2a shows the average density profile and Fig. 2b the standard deviations from it. These are found to be two orders of magnitude smaller than the mean value at each level. When we analyze the relative variation of density standard deviation at each level compared to the averaged value, the fluctuations are typically less than 1% of the mean value, except between 15 and 20 km height (Fig. 2c).

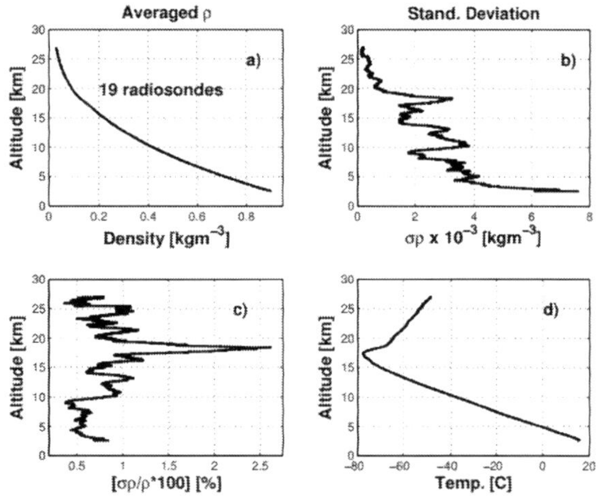

Figure 2. a) Average density profile, b) density standard deviation profile from the mean value, c) density standard deviation relative variation from the mean value and d) averaged temperature profile from radiosondes launched at Paranal.

A closer look at the layer 15-22 km shows that temperature profiles observed during the radiosonde campaign varied almost 20 K close to 18 km (Fig. 3). Therefore, the large relative density fluctuations observed in that layer are due to the large temperature variations that occurred near the tropopause level. The RS92-SGP radiosondes measure temperatures in the range -90°C to +60°C with a resolution of 0.1 °C. The recorded temperatures near the tropopause are inside the technical specifications. However, the horizontal displacement of radiosondes at 18 km ranged between 45 and 100 km, which could be one of the reasons of the large observed temperature variations.

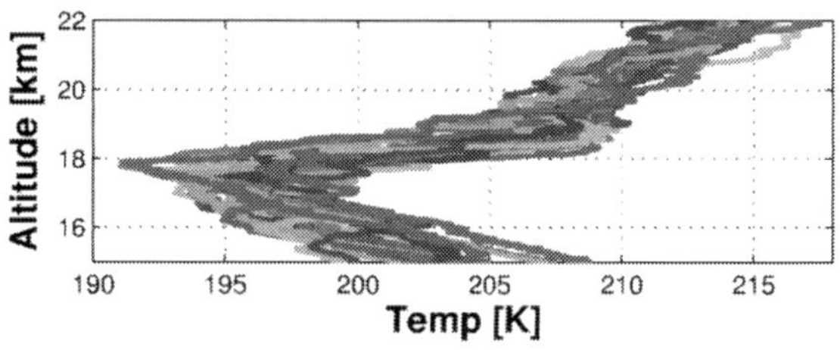

Figure 3. Zoomed view between 15 and 22 km height of temperature profiles measured by 19 radiosondes launched at Paranal during the period of study.

3.2 Comparison between WRF and radiosondes

Density profiles calculated from WRF simulations at the nearest grid-point to Paranal were compared to radiosondes. Just 12 radiosondes were used in the comparison since simulations were not available for some days of the study period. The mean density profile from simulations show close agreement with observations (Fig. 4a) from the surface to 20 km height.

The maximum density relative error between simulations and observations is plotted at each level from the surface to 20 km to analyze the maximum deviation from

DOI: 10.1051/epjconf/201714401007

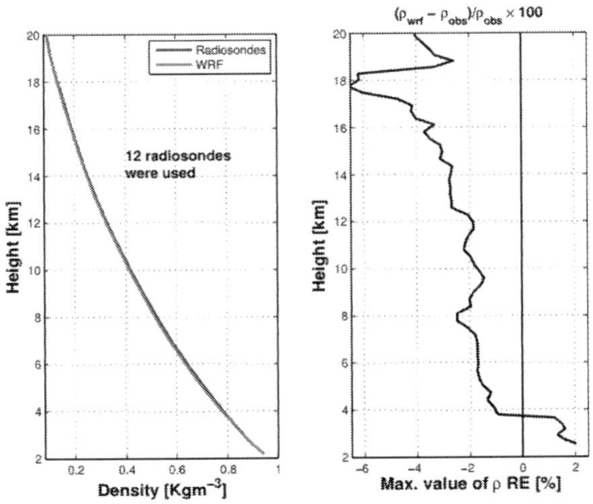

Figure 4. a) Averaged density profiles from WRF simulations and radiosondes and b) maximum density relative error at each level during the period of study.

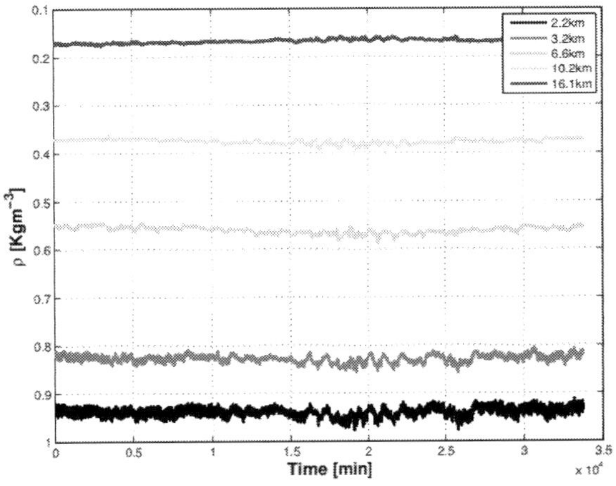

Figure 5. Temporal evolution of simulated atmospheric density from the nearest grid-point to the CTA site at different levels in the atmosphere (see legend for details).

observations that can be expected from WRF simulations (Fig. 4b). The model overestimates the atmospheric density below 4 km height and underestimates it above that altitude. Maximum absolute density relative errors less than 3% are shown in the troposphere. The relative error can increase close to 7% at the tropopause and lower stratosphere (15-19 km height), where the largest density relative variations seems to occur.

3.3 Density seasonal and diurnal evolution

Since the WRF model represents well the observed density profiles at Paranal site, WRF simulations were used to analyze the seasonal and diurnal evolution of density at the CTA site, where no radiosonde measurements are currently available.

The temporal evolution of simulated density in 2011 shows larger absolute variations at lower than at higher levels in the atmosphere (Fig. 5).

The relative density variation $(\rho - \rho(year))/\rho(year) \times 100$ was calculated from WRF simulations (Fig. 6). Its evolution during 2011 shows a clear seasonal trend at all levels in the atmosphere. Density decreases in austral spring and summer months and increases in fall and winter months below the tropopause level. On the other hand, a reverse variation is shown at the highest levels. Density increases in spring and summer months and decreases at fall and winter months. The largest density relative variations occurs near the tropopause and lower stratosphere. At those heights, large density relative variations can be seen in short periods of time, particularly during winter and spring months.

We grouped the simulated density profiles by season and calculated its relative variation over the year $((\rho(season) - \rho(year))/\rho(year) \times 100)$. The seasonal density profile varies less than 1% relative to the annual mean within the troposphere but in the tropopause and lower stratosphere the seasonal density variation is larger than 3% of the annual mean (Fig. 7). The largest density relative variations occur in summer and winter months.

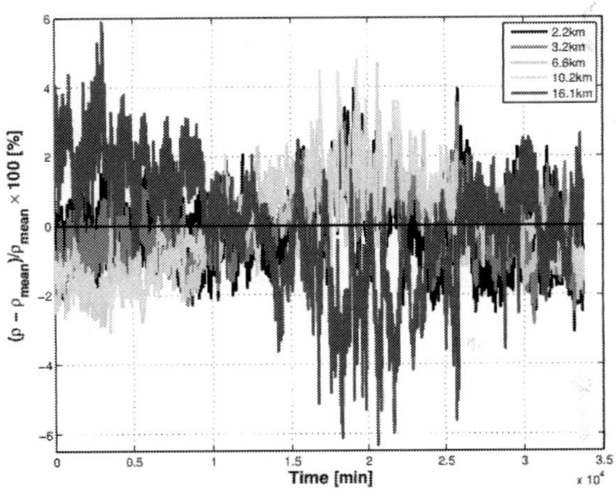

Figure 6. Temporal evolution of relative density variation $(\rho - \rho(mean))/\rho(mean)$ from WRF simulations at the nearest grid-point to the CTA site at different levels in the atmosphere (see legend for details).

The diurnal evolution of density was studied by calculating the relative density variation at each hour compared with the annual mean. Fig. 8 shows that the largest density relative variations occur near the surface and the lower stratosphere during the day. The lowest density relative variations are shown during afternoon hours and the largest density relative variations are shown between 2-8 h local time.

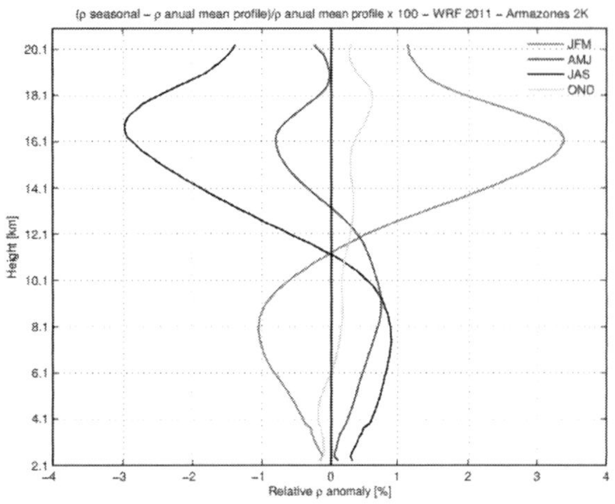

Figure 7. Relative seasonal density deviations from annual mean profiles from WRF simulations at the nearest grid-point to the CTA site (see legend for details).

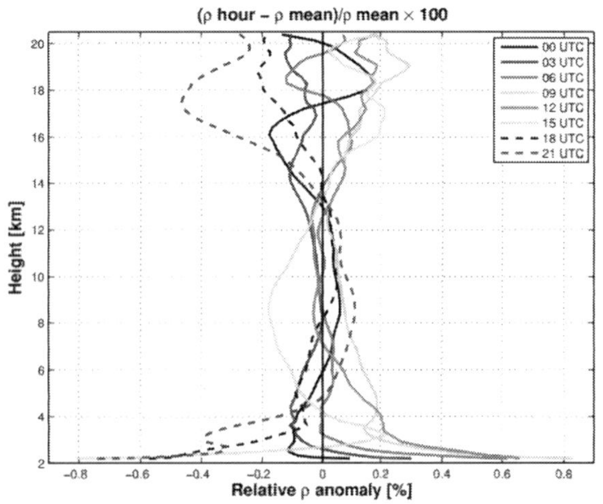

Figure 8. Three-hourly relative density deviations from annual mean profiles from WRF simulations at the nearest grid-point to the CTA site (see legend for details).

4 Conclusions

The Cherenkov Telescope Array (CTA) in the southern hemisphere will be installed at Armazones 2k site in northern Chile. Scarce atmospheric observations are available in the region, particularly radiosonde data. However, a small radiosonde campaign was conducted at Paranal observatory, located at about 21 km from the CTA site, from 24 October and 4 November 2011.

Radiosonde observations at Paranal showed that the largest density variability occurs near the tropopause, as a result of temperature changes. Figure 1b of [9] also showed the largest variations in density profiles to occur between 10 and 20 km depending on location and season.

High-resolution numerical simulations with the Weather Research and Forecasting (WRF) model were performed almost every day in 2011 to provide weather forecasts for the region of study. Simulations were compared with radiosondes launched at Paranal observatory. Maximum relative errors between WRF and observations were less than 2% in the troposphere but slightly increase at the tropopause and lower stratosphere. Despite the good agreement between the model and observations, more observations are needed to have more confidence on the model results in order to use it in the region.

The simulated density is used to study its seasonal and diurnal variation at the CTA site. Absolute variations at Armazones 2k site are larger near the surface than at higher levels. In addition, density relative variations are larger in the lower stratosphere.

Density relative variations in the lower stratosphere increase in summer (up to 5%) and during the day (up to 0.3%) and decrease in winter (up to -6%) and during afternoon hours (up to -0.45%) at Armazones 2k. The simulated density relative variations near the surface increase in winter (up to 4%) and near dawn (up to 0.82%) and decrease in summer (about -3 %) and during afternoon hours (about -0.8%) at Armazones 2k.

Acknowledgements

We acknowledge the European Southern Observatory (ESO) and Paranal staff for its support during the radiosonde campaign at Paranal site. Global Forecast System (GFS) model forecasts provided the initial and boundary conditions for the WRF model and were downloaded from the National Centers for Environmental Prediction (NCEP) servers. We acknowledge the Centro de Estudios Atmosféricos y Astroestadística (CEAAS) from Universidad de Valparaíso.

References

[1] https://www.mpi-hd.mpg.de/hfm/HESS/
[2] http://veritas.sao.arizona.edu/
[3] https://magic.mppmu.mpg.de/
[4] Actis, M., et al., Experimental Astronomy, 32, 193–316 (2011)
[5] Bernlöhr, K, Astroparticle Physics, 12, 255-268 (2000).
[6] Louedec, K., and M. Will, J. Phys.: Conf. Ser. 409, 1, (2013).
[7] Abreu, P., et al., Astroparticle Physics, 35, 9, 591–607 (2012)
[8] Skamarock WC, et al., NCAR/TN-468+STR NCAR TECHNICAL NOTE (2008)
[9] Daniel, M.K., Journal of Physics: Conference Series, 595 (2015)

DOI: 10.1051/epjconf/201714401008

Raman LIDAR for UHECR experiments: an overview of the L'Aquila (Italy) lidar station experience for the retrieval of quality-assured data

Marco Iarlori[1,*], *Vincenzo* Rizi[1], *Giuseppe* D'Amico[2], *Volker* Freudenthaler[3], *Ulla* Wandinger[4], and *Aurelio* Grillo[1]

[1]CETEMPS and INFN, Dipartimento di Scienze Fisiche e Chimiche, Università degli Studi dell'Aquila, via Vetoio, 67100, L'Aquila, Italy
[2]Consiglio Nazionale delle Ricerche – Istituto di Metodologie per l'Analisi Ambientale (CNR-IMAA), Contrada S. Loja – Zona Industriale, 85050 Tito Scalo (Potenza), Italy
[3]Ludwig-Maximilians-Universität, Meteorologisches Institut Experimentelle Meteorologie, Munich, Germany
[4]Leibniz Institute for Tropospheric Research, Permoserstr. 15, 04318 Leipzig, Germany

Abstract. L'Aquila (Italy) lidar station is part of the EARLINET (European Aerosol Research Lidar Network) since its beginning in the 2000. In the EARLINET community great efforts are devoted to the quality-assurance of the aerosol optical properties inserted in the database. To this end, each lidar station performed intercomparisons with reference instruments, a series of internal hardware checks in order to assess the quality of their instruments and exercises to test the algorithms used to retrieve the aerosol optical parameters. In this paper we give an overview of our experience within EARLINET quality-assurance (QA) program, which was adopted for the Raman lidar (RL) operated in the AUGER Observatory. This program could be systematically adopted for the lidar systems needed for the current and upcoming UHECR experiments, like CTA (Cherenkov Telescope Array).

1 Introduction

The EARLINET (European Aerosol Research Lidar Network, www.earlinet.org) was founded in year 2000, involving 19 lidar systems located in 11 countries, with the main purpose of establishing an aerosol climatology for Europe. EARLINET community has grown and it currently consists of 35 lidar systems in 16 countries [1]. At beginning, the EARLINET lidar systems consisted mainly in single wavelength UV Raman lidars, like the one that is operated in L'Aquila [2], while nowadays a lot of them have been upgraded to the so-called 3+2+δ Raman lidars. The expression 3+2+δ means that these modern Raman lidars are able to retrieve independently the profiles of the backscatter coefficients at 355 nm, 532 nm and 1064 nm wavelengths, and of 2 aerosol extinction coefficients at 355 nm and 532 nm, and of the aerosol linear depolarization ratio [3]. The multi-wavelength capability allows to retrieve the aerosol microphysical properties as well [4].

The lidar systems in EARLINET are very different each other being the instruments designed and assembled in-house; the same is true for the retrieval algorithms, anyway the EARLINET community is making significant improvements to uniform the retrieval of the aerosol optical parameter profiles developing the Single Calculus Chain [5].

The different lidar systems are continuously upgraded, and for these reasons, the EARLINET community adopted an intensive quality-assurance (QA) program. That QA program was improved in the successive collaborations like EARLINET–ASOS (Advanced Sustainable Observation System) and ACTRIS (Aerosols, Clouds, and Trace gases Research InfraStructure), and it is still operative in the framework of ACTRIS-2 research infrastructure project.

We briefly describe the EARLINET QA program and, as an example, some of the results related to L'Aquila RL.

Finally, we propose to adopt the QA program activities for the lidar systems needed to UHECR experiments.

2 The EARLINET QA program

The EARLINET QA program at instrument level consists of:

- intercomparison campaigns with a reference lidar system [6];
- applying internal system check-up tools, developed to perform tests on regular basis and/or each time the lidar system is substantially changed [7].

At software level, several exercises have been prepared to test:

* Corresponding author: marco.iarlori@aquila.infn.it

 DOI: 10.1051/epjconf/201714401008

- the algorithm to retrieve the profile of the backscatter coefficient from elastic-backscatter lidar signals [8];
- the algorithm to retrieve the profiles of the backscatter and extinction coefficient from RL signals [9].

2.1. Lidar systems Intercomparison

The first EARLINET lidar intercomparison campaign was performed in 2000 and was based on a previous successful experience in the framework of the German Aerosol Lidar Network [10]. In this campaign it was found that, among the lidar systems, the typical mean deviations of aerosol backscatter coefficients in both the planetary boundary layer and free troposphere were within the predefined quality criteria [11, 6].

Other campaigns were performed by comparing the data taken with each lidar system to the ones simultaneously measured with a co-located reference lidar.

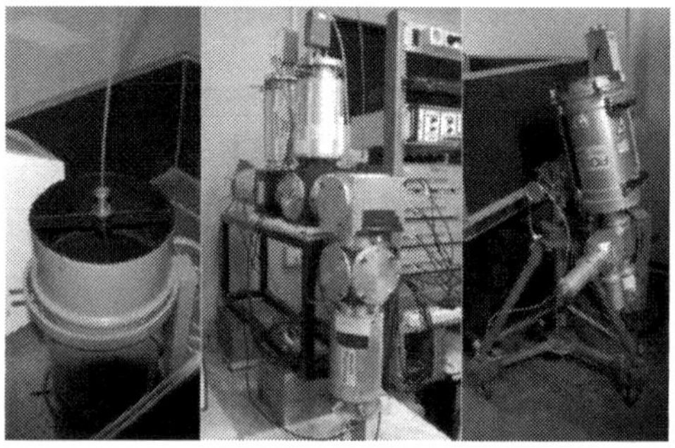

Fig. 1. Left and central panel: the L'Aquila RL receiver. Right panel: the reference lidar system for the 2012 intercomparison campaign, i.e. the POLIS (Portable Lidar System) system of the Ludwig-Maximilians-Universität, Meteorological Institute, Munich, Germany.

Because of the substantial upgrade of the lidar systems, the most recent intercomparison campaigns involved both elastic and Raman channels and, as consequence, the comparison of the aerosol backscatter and extinction profiles. The raw lidar signals were also compared to check for system problems, their causes and possible solutions [6].

These recent campaigns evidenced that most of the EARLINET lidar systems performs well, for example, the mean relative systematic signal deviations from the reference lidar system were found to be within 1–2%. The comparison of the aerosol optical properties has shown that the aerosol backscatter could be retrieved with an accuracy better than $\pm 2\times10^{-4}$ km^{-1} sr^{-1} and the extinction better than ± 0.01 km^{-1}.

The L'Aquila lidar system performed quite well in all the intercomparison campaigns. For this reason, it was used as reference for the Auger Observatory RL [12].

2.2 Hardware check-up tools

Part of the instrumental QA is constituted by a number of tools. These tools are useful to test the performances of lidar sub-systems, i.e. telescope–laser alignment. EARLINET lidar stations have to perform these tests on regular basis, These tools are fully documented [7] and we give only a brief overview of their use.

2.2.1 Telecover test

There is no calibration method for a lidar system in the near range, where clean air conditions are almost never met. The near range is also where the receiver optical performances could have the greatest impact.

In the telecover test [13], the backscattered photons collected from different sectors of the telescope are compared. In order to perform these measurements, the telescope aperture is covered leaving open only a defined sector at time (north (N), south (S), east (E) and west (W) quadrants, see Fig. 2). In the ideal case these signals have to exhibits no difference, but the overlap range. Measurements with full telescope aperture are done to check if the atmospheric condition can be considered stable during the test.

Fig. 2. On the left panel it is shown the nomenclature of the quadrants for the telecover test. On the right panel, the primary mirror during the test.

If range dependent differences are spotted, a further inspection to find out the possible causes should be made (for example misalignments or spatial inhomogeneity of the detector sensitivity, should be investigated). It could be also useful to have a ray tracing simulations of the lidar receiver.

The mean signal (M) is the average of the signals returns in each sector: M=(N+E+S+W)/4, then the relative deviations could be calculated as: Ndev = (N – M)/M for each sector. The overall relative deviation is: AllDev = sqrt [(Ndev^2 +Edev^2 + Sdev^2 + Wdev^2)/4].

Internal quality criteria [7] related to the values of the relative deviations help to qualify the alignment and the optical performances of the lidar system. L'Aquila RL has fulfilled these criteria along all the performed tests.

2.2.2 Rayleigh fit test

While the telecover test checks the lidar system performance in the near range, the Rayleigh fit test is the check-up tool for the system behaviour in the far range i.e. in the free troposphere and beyond. In these regions

DOI: 10.1051/epjconf/201714401008

are likely the presence of aerosol-free atmospheric layers and is possible to measure the agreement between a calculated pure molecular signal and a real lidar signal.

Basically, the Rayleigh fit is a normalization of a lidar signal to the calculated attenuated Rayleigh backscatter coefficient in an aerosol free region. EARLINET internal criteria for the quality of the fit are based on statistical analysis of the fit residuals [7]. An evaluation "by eye" of the fit by a lidar expert is sufficient to recognize the more evident signal distortions, not reported during the tests on the L'Aquila RL, and then investigate for possible causes (e.g. misalignment).

2.2.3 Zero-bin test

If the delay between the laser pulse emission and the start of the acquisition system is not well known, or in other words, if the range scale is not correctly set, there could be a quite large indetermination in the retrieval of the aerosol extinction in the near range, especially in the very first kilometer. This range dependent absolute value of the extinction indetermination ($\Delta\alpha$) could be calculated as function of the uncertainty (r_0) in the position of the true zero-range [7]:

$$\Delta\alpha(r, r_0) = \frac{2}{1+\left(\dfrac{\lambda_{Raman}}{\lambda_{Laser}}\right)^k} \frac{1}{(r-r_0)^2} r_0 \cdot \qquad (1)$$

In Eq.(1), r is the range, λ_{Laser} is the laser wavelength, λ_{Raman} is the wavelength of the N_2 Raman backscattered photons, and k is the Ångström exponent, which depend mainly on the aerosol size and it could be set to 1 for most of the aerosol in the planetary boundary layer (PBL). In this condition, if the resolution of the acquisition system is 15 m and the zero-range uncertainty r_0 is of just one range bin, the absolute value $\Delta\alpha$ at 355 nm is about 0.1 km^{-1} at 0.4 km range and it became < 0.01 km^{-1} only above 1 km range. Then also for a relatively small r_0 value, $\Delta\alpha$ could be of the same order of the typical extinction value in the PBL.

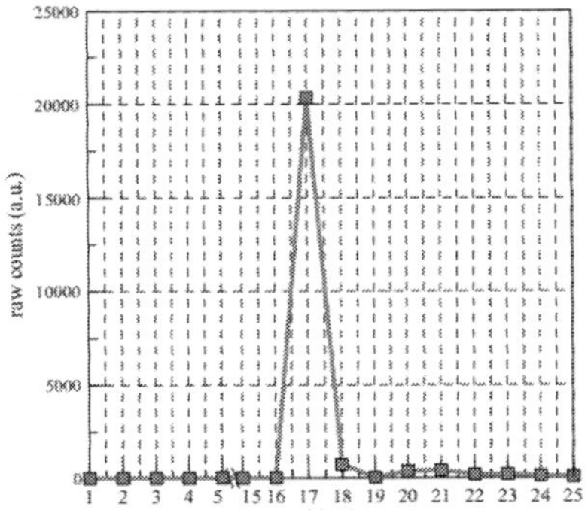

Fig. 3. Zero-bin measurement for the L'Aquila UV RL. Pre-triggering allows to measure the first range gate associated to the start of the laser pulse (the 17th bin).

Pre-trigger are often set in the acquisition of a lidar system, i.e. the acquisition starts to record before the actual time of the laser pulse emission. In this condition the zero-bin test can be performed: the lidar system can measure the photons reflected by a nearby obstacle (i.e., into laboratory). A quite definite peak is usually recorded, and it marks the location of the zero-bin, i.e. the actual start of the range scale (Fig.3).
This test has been performed on the RL build for the Auger Observatory [12]. Because the UHECR measurement sites are particularly clean from point of view of the aerosols content, an accurate zero-bin measurement is important to decrease the magnitude of the systematic uncertainty in the aerosol optical depth estimation.

2.2.4 Dark measurement

A dark measurement is a standard lidar measurement performed with the telescope fully covered. In this way the signal acquired will contain no atmospheric backscattered photons, but only a signal due to interference sources, for example the various spurious pick-ups or other EM interferences synchronized with the laser trigger or any other instrumental distortion.

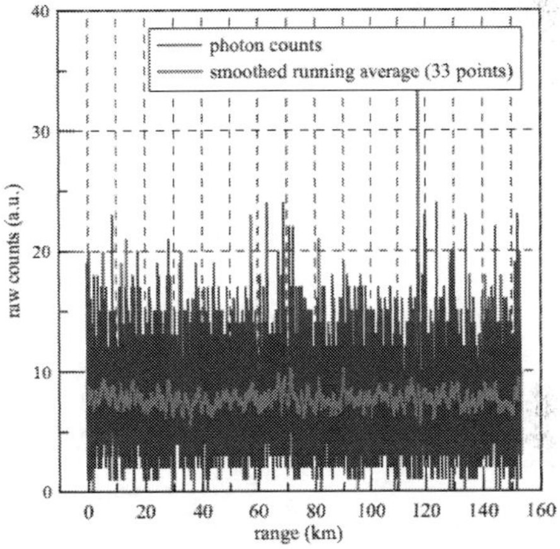

Fig. 4. An example of the dark measurement performed on the L'Aquila UV RL. It does not evidence any particular behavior.

Although currently there is no a standardized way to deal with the peculiar and not-stationary distortions that could be evidenced by this measurements, they are anyhow valuable because capable to point out the existence of hardware problems. The L'Aquila RL dark measurements do not show any distortions (Fig.4).

2.3 Algorithms Intercomparison

In the framework of EARLINET, several exercises have been carried out to ensure the quality of the algorithms for the retrieval of the aerosol optical parameter profiles. These tests on the algorithms were performed since the very start of the collaboration and all the results and the details can be found elsewhere [8, 9].

DOI: 10.1051/epjconf/201714401008

2.3.1 The aerosol backscatter inversion algorithm test (single elastic-backscatter lidar signal)

The aim of this test was to investigate the implementation of the so-called Fernald inversion scheme [14] (or equivalent methods) in the algorithms used in each of the EARLINET stations. To this end, simulated elastic lidar signals were provided to the groups.

This blind test were organized in three different stages and each stage had several cases. In the first stage, the only additional information available was the atmospheric molecular density profile. In the second stage, the lidar ratio profile was provided and finally also the calibration parameters were made available. The first two stages provide the sensitivity to the input parameters, while the third stage was useful to actually test the goodness of the algorithms. The test was performed with simulated signals at 355nm, 532nm and at 1064nm.

The test evidenced the needs of an accurate estimation of the lidar ratio profile in order to retrieve a meaningful aerosol backscatter profile, especially for the UV signal. On the other hand, the retrieval at longer wavelengths evidenced stronger sensitivity to the calibration parameters.

The results for the third stage shown that all the EARLINET stations have a suitable algorithms for the retrieval of the aerosol backscatter profile.

This was also the case for the elastic-backscatter lidar inversion algorithm developed in L'Aquila, that in the third stage of the intercomparison produced profiles with an overall relative deviation from the true profile <1% at all wavelengths [8].

2.3.2 The aerosol backscatter and extinction inversion algorithm test (Raman lidar signals)

The RL independently measures the aerosol extinction and backscatter profiles [15]. A non-scanning elastic lidar system is unable to do the same [16].

The retrieval algorithm for the extinction, has its key part in the numerical estimation of a derivative of the Raman return [9]; there are several methods to perform it, and most of them involve the use of smoothing filters. As a consequence of the filtering operation on the data, the range resolution of retrieved products become degraded [17].

The test of the Raman algorithms used in each of the EARLINET groups was blind and involved the algorithm for retrieval of the aerosol extinction profile starting from Nitrogen Raman signals, the algorithm for the retrieval of the aerosol backscatter and the lidar ratio estimation. The simulated Raman and elastic lidar signals were produced with a realistic aerosol profiles and noise.

The result of the algorithms tests are reported in Pappalardo et al. [9]. The mean deviation from the solution, for both the extinction and the backscatter profiles, have been found within the limits fixed in EARLINET [11]; for the lidar ratio, the mean relative deviations were found compatible with signal to noise ratio (SNR). The estimations of the statistical errors have been analyzed and they were consistent with the SNR embedded in the synthetic signals.

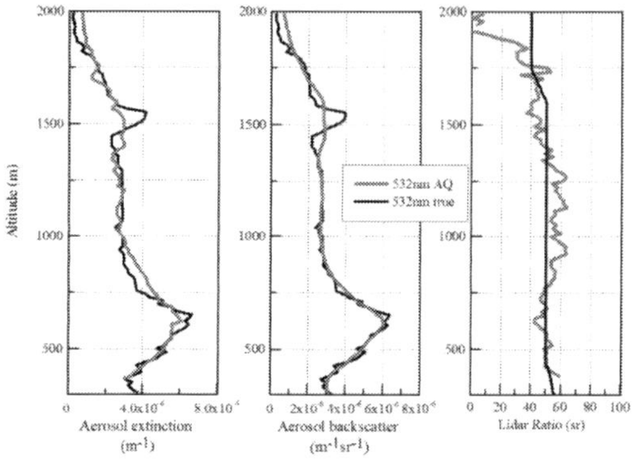

Fig. 5. The Raman algorithm test at 532nm for the L'Aquila lidar station. Beside the general agreement between the true profiles (black lines) and the retrieved profiles (red lines), it can be noted how the smoothing filter embedded in the extinction retrieval algorithm, removed the peak at 1500m.

All the algorithms performed quite well, including the ones developed in L'Aquila, as it can be seen in the example reported in Fig.5. This figure also shows the importance of a correct use of the smoothing in order to get a consistent lidar ratio [17].

3 Conclusions

The key guidelines and methods of the intensive QA program developed by the EARLINET lidar community, have been briefly overviewed. The quality assurance is thought to be a dynamic process always ongoing, not just a one-time exercise.

It seems recommendable to adopt the ideas and concepts of the EARLINET QA program for the lidars built to be employed in UHECR experiments, in order to have data whose quality could be constantly traced.

As part of EARLINET, the L'Aquila RL has been successfully tested in hardware intercomparison campaigns, by continuative use of the internal hardware check-up tools, and in the quality checks of the inversion algorithms .

The L'Aquila lidar station, although is not one of the EARLINET reference systems, could be used as benchmarking system for lidars dedicated to the UHECR experiments. L'Aquila could also be considered as a source of certified algorithm codes for the retrieval of the profiles of the aerosol optical parameters.

Acknowledgements. The financial support for EARLINET in the ACTRIS Research Infrastructure Project by the European Union's Horizon 2020 research and innovation programme under grant agreement no. 654169 and previously under grant agreement no. 262254 in the 7th Framework Programme (FP7/2007–2013) is gratefully acknowledged.

References

1. G. Pappalardo, et al., Atmos. Meas. Tech., **7**, 2389–2409, doi:10.5194/amt-7-2389-2014 (2014).
2. V. Rizi, et al., Appl. Opt. 43, 6440-6453 (2004).
3. V. Freudenthaler, et al., Tellus B, **61**, 165–179, doi: 10.1111/j.1600-0889.2008.00396.x, (2009).
4. D. Müller, et al., Atmos. Meas. Tech., **9**, 5007–5035, doi: 10.5194/amt-9-5007-2016 (2016).
5. G. D'Amico, et al., Atmos. Meas. Tech., **8**, 4891–4916, doi: 10.5194/amt-8-4891-2015 (2015).
6. U. Wandinger, et al., Atmos. Meas. Tech., **9**, 1001–1023, doi: 10.5194/amt-9-1001-2016 (2016).
7. V. Freudenthaler, et al., Atmos. Meas. Tech. Discuss., 2017 (in preparation).
8. C. Böckmann, et al., Appl. Opt. **43,** 977–989 (2004).
9. G. Pappalardo, et al., Appl. Opt., **43**, 5370–5385 (2004).
10. J. Bösenberg, et al., Rep. 317 Max-Planck-Institut für Meteorologie, Hamburg, Germany (2001).
11. V. Matthias, et al., Appl. Opt. **43,** 961–976 (2004).
12. L. Wiencke, et al., *Joint Elastic Side-Scattering Lidar and Raman Lidar Measurements of Aerosol Optical Properties in South East Colorado*, submitted to J. Instrum.
13. V. Freudenthaler, *The telecover test: a quality assurance tool for the optical part of a lidar system.* Reviewed and Revised Papers Presented at the 24th International Laser Radar Conference, Boulder, Colorado, USA, 23–27 June 2008, Boulder, Colorado, USA, presentation: S01P-30, available at: http://epub.ub. uni-muenchen.de/12958/ (2008) (last access: 10 November 2016).
14. F. G. Fernald, Appl. Opt. **23**, 652– 653 (1984).
15. A. Ansmann, et al., Appl. Opt. **31**, 7113–7131 (1992).
16. R. T. H. Collis and P. B. Russell, in *Laser Monitoring of the Atmosphere* (Springer-Verlag, Berlin, 1976) pp. 71–151.
17. M. Iarlori, et al., Atmos. Meas. Tech., **8**, 5157–5176, doi: 10.5194/amt-8-5157-2015 (2015).

EPJ Web of Conferences **144**, 01009 (2017) DOI: 10.1051/epjconf/201714401009
AtmoHEAD 2016

Searching for Anomalous Longitudinal Profiles with the FRAM Telescope

Jiri Blazek[1]* for the Pierre Auger Collaboration[2]**,***

[1] *Institute of Physics, Czech Academy of Sciences, Prague*
[2] *Observatorio Pierre Auger, Av. San Martín Norte 304, 5613 Malargüe, Argentina*

Abstract. The study of anomalous longitudinal profiles of extensive air showers presents an interesting opportunity to gain additional insight about the character of hadronic interactions at the highest energies. Moreover, the presence of such profiles would represent direct evidence of a light component in the primary cosmic rays of a given energy, independently of hadronic interaction models. The ratio of profiles observed to have a "double bump" structure (with two clearly distinguishable atmospheric depths of maximum particle count) compared to the total number of events can moreover be used to test the predictions of interaction models. However, the majority of such profiles observed at the Pierre Auger Observatory are caused by clouds. Here we describe a method of rapid monitoring of particularly interesting cosmic ray events, using the F/Photometric Rapid Atmospheric Monitor (FRAM) telescope to identify events with clear, cloudless atmospheric background. The function of FRAM is described in detail and the number of triggered events is presented.

1 Introduction

The study of the ultra-high energy cosmic rays (UHECRs), usually denoting primary particles with energies higher than 10^{18} eV, is uniquely positioned to probe hadronic interactions in an energy range well beyond the capabilities of the current generation of colliders. Another interesting and open question is that of the chemical composition of the incoming primary charged particles, see, e.g., [2]. Studies of this type focus on the average features of the longitudinal profile of a shower, such as the mean slant depth of its maximum particle count and the mean variance of this value, as recorded with the nitrogen fluorescence technique. Recent results from the Pierre Auger Observatory [1] indicate a mixed composition which is becoming heavier with energy [3], but the interpretation of the relevant measurements is burdened with large systematic uncertainties related to differences between predictions of various hadronic interaction models. Several other observables sensitive to the chemical composition are being investigated, such as the asymmetry in the rise-time of the signal of the surface detectors [4]. But another such observable should in principle be measurable: assuming a light component in the chemical composition, some of the observed longitudinal profiles should feature a distinct second maximum. Such a profile is created by a highly energetic and deeply penetrating particle originating during the first few generations of a shower development. The position of the second maximum and the rate of occurrence of such "anomalous" showers are sensitive both to the chemical composition of the primary flux and to the properties of the hadronic interaction models such as the cross section, elasticity and multiplicity, which are extrapolated from the latest collider data.

The longitudinal profiles are generally very well described by the so called Gaisser-Hillas (G.-H.) function [5], an empirical Ansatz of the form

$$N_{GH}(X) = N_{max}\left(\frac{X - X_0}{X_{max} - X_0}\right)^{\frac{X_{max}-X_0}{\lambda}} \exp\left(\frac{X_{max} - X}{\lambda}\right),$$

where X is the atmospheric slant depth, N_{GH} the number of particles at that given depth, N_{max} the number of particles at the shower maximum depth X_{max} and the remaining X_0 and λ are energy-dependent fit parameters without a consistent physical interpretation. The latter two correspond roughly to the depth of a first interaction point and a radiation length, respectively, but the best fit often assigns them unphysical values.

A small fraction ($\sim 10^{-3}$) of both the observed and simulated showers features a deeply propagating primary particle, causing the location of a shower maximum to be significantly deeper in the atmosphere than the mean X_{max} for a given particle type. A similar phenomenon can also occur during the development of a single shower, when a (leading) particle leaves the evolving cascade early during shower development and manages to propagate deeply into the atmosphere. Such a particle then forms a secondary sub-shower, which, if it possesses a significant fraction of the energy of the primary particle, is clearly discernible from the primary sub-shower. The occurrence of such an event is highly unlikely in the case of heavy nuclei acting as primary particles, due to simple kinematic reasons. An Fe-56 nucleus, in the crude approx-

*e-mail: blazekj@fzu.cz
**e-mail: auger_spokespersons@fnal.gov
***Author list: www.auger.org/archive/authors_2016_09.html

DOI: 10.1051/epjconf/201714401009

imation of a superposition model, can be seen as a system of 56 independently interacting nucleons, each possessing approximately 2% of the total energy. A clearly identified anomalous profile would then present direct evidence for a light component (protons) at a given energy. Clearly, the anomalous profile should be well described by a combination of two regular Gaisser - Hillas profiles $N_{DGH}(X) = N^1_{GH}(X) + N^2_{GH}(X)$.

2 Monte Carlo Study

The rate of occurrence of anomalous showers can be examined by looking at the properties of a statistically large sample of simulated longitudinal profiles. For this kind of study, where the lateral features and the distribution of particles at the ground are of no interest, the CONEX [6] software presents an ideal tool. A range of energy of $10^{17} - 10^{20}$ eV is investigated, divided into seven bins, each with approximately 10^5 simulated events. The EPOS-LHC interaction model is used, the primary particle being a proton incident at a zenith angle of 60°. Three independent methods for the identification of anomalous profiles are devised: one looking at the number of inflexion points present in the profile, the second one making use of existing ROOT routines for extracting signal from background and the third routine discriminating the anomalous profiles based on the quality of the fit of the $N_{DGH}(X)$ in comparison to the $N_{GH}(X)$. Results from the first two methods are shown in figures 1 and 2, respectively. Each method depends on one or several parameters that define the double-bump discrimination, three choices of such parameters being shown in each figure. The outcomes of the third method are consistent with the results shown in Figs. 1 and 2. The fraction of anomalous profiles is of the order of 10^{-3} and decreases with energy at first, which is expected since the cross section rises with energy [3] and therefore the two maxima become less distinguishable. The origin of the rise in the highest energy bins is not clear and perhaps warrants further study. However this feature is unlikely to be observable with sufficient statistics due to the steeply falling flux of the primary particles at this energy. Qualitatively similar results were previously obtained by C. Baus et al. [7], but with the notable differences of employing pre-LHC era interaction models and smearing the simulated longitudinal profiles to account for the variance present in real fluorescence detectors.

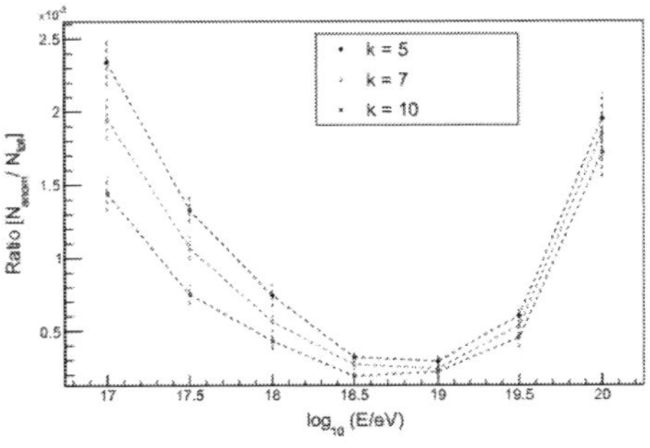

Figure 1. The ratio of identified anomalous profiles N_{anom} compared to the total number of simulated events N_{tot}, based on the method of identifying the number of inflexion points. The method depends on a parameter k, which fixes the minimal separation of two neighboring inflexion points in terms of the number of bins of slant depth. Therefore it also fixes the minimal width of both identified peaks and thus using a higher value of k rejects a set of events corresponding to thinner sub-showers. Three plausible choices of k are shown.

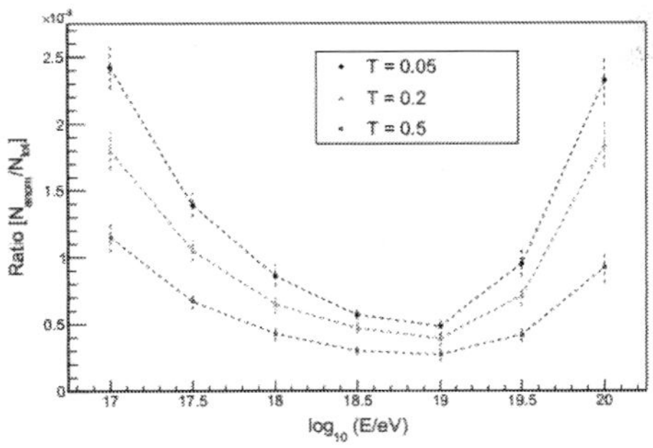

Figure 2. The ratio of identified anomalous profiles based on the method of extracting a number of peaks from background. The method depends on several parameters, the most impactful parameter being the threshold value T, which roughly corresponds to the minimum ratio between the amplitudes of the secondary and the primary peak. Three plausible choices of T are shown.

3 Shoot-the-Shower (StS) Program at the Pierre Auger Observatory

The largest obstacle in studying the anomalous profiles is the fact that the majority of "bumpy" profiles recorded by real fluorescence detectors are caused by effects of atmospheric origin. There are two main mechanisms that can create an artificial anomalous profile. The first one is the absorption of light between the shower axis and the detector by an obfuscating atmospheric layer - typically a cloud - which consequently can create an apparent minimum in the observed profile shape. The second mechanism is the scattering of Cherenkov light when the shower happens to pass through a cloud, creating a second, false peak in the detector due to increased light below the cloud. A proper identification of the presence of such light-affecting layers for each candidate shower is therefore of principal importance. An attempt to address this issue at the Pierre Auger Observatory is the so-called Shoot-the-Shower program. It aims to rapidly record the atmospheric conditions in the vicinity of the axis of a recently recorded shower. It is worth noting that at present there are several types of triggers for the Shoot-the-Shower program, such as particularly-well reconstructed showers, high-energy showers and photon candidates, though the majority of recorded events are still caused by triggers on anomalous profiles. An example of an anomalous event recorded at the Los Leones site of the Pierre Auger Observatory is depicted in Fig. 3. The sharp peak is likely caused by the scattering of Cherenkov light on clouds. The FRAM telescope (described below in the text) does not have information about this particular event.

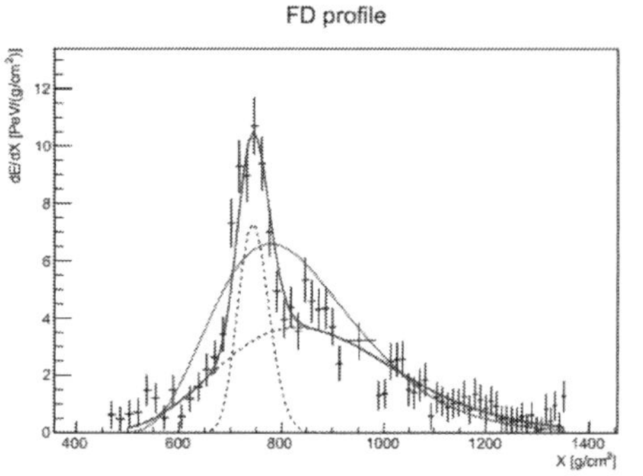

Figure 3. Example of an anomalous fluorescence profile. The red line represents a single G.-H. fit, dashed blue lines represent the fitted sub-showers during a double G.-H. fit and the solid blue line is their combination.

3.1 The FRAM Telescope

Several approaches for rapid atmospheric monitoring were tried in the past at the Pierre Auger Observatory. In addition to launching atmospheric balloons after arrival of a very high energy shower to describe actual atmospheric profiles [8], the lidars [9] have been used to obtain the cloud information. The current and running implementation makes use of the so called F/Photometric Rapid Atmosphere Monitor - FRAM [10], located at the Los Leones site. Its main advantage is its completely automatic and autonomous functionality and fully passive operation, which means that it does not interfere with the regular measurements of the fluorescence detectors at all. The working mechanism of FRAM is in principle very simple - it records a series of images in the vicinity of a shower axis, finds stars in these images and compares their apparent brightness with the catalogued values. The FRAM setup consists of several astronomical devices, of which the wide-field camera (field of view 7° x 7° fixed on a remotely controlled robotic mount) is used for the StS program. When receiving the trigger information, the camera shoots a series of images along the shower axis obtained by an online reconstruction. This whole process takes 5 minutes on average. The images are then copied to a remote storage site and analyzed at a later time.

A first step in the image processing is the application of a dark-field image to correct for the background variance due to any dark current present in the camera system and then the application of a flat-field image to partly correct the optical imperfections of the camera such as vignetting. The resulting calibrated image is then run through an astrometry software suite which identifies potential sources in the recorded field and finds a precise location on the sky. For each identified source, the value of the recorded signal is measured using aperture photometry; the brightest stars are cut from the data-set due to saturation. The remaining sources with apparent magnitudes corrected for the amount of average airmass in the atmosphere along the line of sight are then matched to stars present in the catalog Tycho-2. If then, on a given portion of the image, a region of sources is well-matched to catalogued stars in terms of position but its apparent brightness is significantly lower, it can be inferred that an obfuscating layer was present in that part of the image.

This whole process can be made visually much clearer when plotting all the images from a given shower scan into one picture, thus covering a wide range of airmasses. A theoretical fit of the expected extinction can be made and sets of stars differing significantly and uniformly from the curve of the fit mark regions with light-affecting layers present. Examples of such plots can be seen in Figs. 4 and 5. In the first figure, there are three visible regions with significant variations from the expected fit, at around 7, 13 and 20 degrees of altitude. The latter two regions represent clouds with an average optical thickness of around 0.3 (denoted as τ and defined as $N_{apparent}/N_{expected} = e^{-\tau}$), which in the linear approximation thus represents roughly 30% attenuation in terms of the absolute photon flux. The presented method is sensitive to minimal optical depths of approximately 0.2. The presence of such regions of significant attenuation would cause the corresponding shower to be discarded for the purposes of the double-bump analysis. Fig. 5 shows a significantly cloudier profile, but note that the stars, though strongly attenuated, are still visible. In many of the recorded images there are wide regions with no discernible stars and then the analysis procedure fails during the first step of astrometry. Such profiles are typical for nights with heavy cloud coverage.

For purposes of the double-bump analysis, the profiles are inspected visually by a member of the FRAM team and only the events with no obvious deviations from the extinction fit are selected and tagged as having a clear atmospheric background. A series of cuts is then applied to the reconstructed fluorescence profiles in order to try to identify the actual anomalous events amid the much larger set of events caused simply by the intrinsic random variations in the corresponding fluorescence profiles. The set of cuts includes the requirement of a significant improvement of the double G.-H. fit over a single G.-H. fit, both maxima being present in the field of view, a minimal track length between the two fitted maxima and other cuts.

3.2 Number of StS Events Recorded by FRAM

The FRAM telescope was originally installed in 2005 and has undergone many hardware improvements. The current wide-field camera, producing best results in the StS program, was installed in March 2013. In this analysis the dataset starting from March 2013 is presented. The dataset extends up to April 2016, with newer data still waiting to be analyzed. During this roughly three-year period, the FRAM telescope registered 1788 events with at least 4 pictures per scan recorded ("complete events"). Of these, 659 were identified as showers with a clear atmospheric background. Of this number, 399 events were shot after receiving a double-bump trigger. These numbers

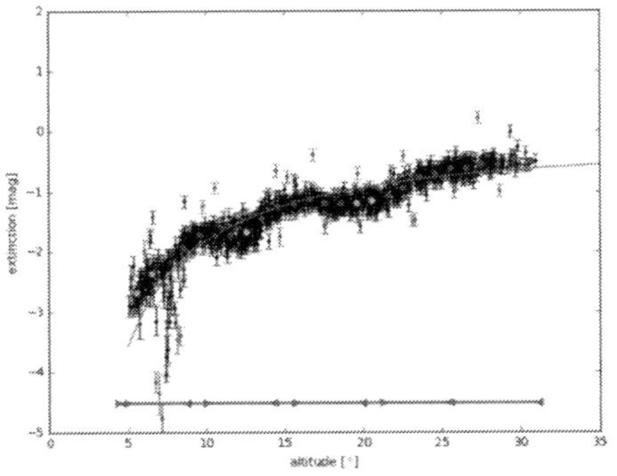

Figure 4. Example of an extinction plot for a FRAM event. The quantity shown on the y-axis is the extinction in magnitudes, corresponding to the combination of aerosol and molecular scattering. On the x axis is the altitude, i.e. the angle between the horizon and the line of sight. Each black dot represents an identified light source. The red line is a fit of total extinction due to molecular and aerosol scattering. An ideal cloudless event would have all points clustered along this line, as the attenuation in such a case depends only on the amount of air in the direction of the star. The airmass, and hence also the attenuation, increases towards the horizon. The large dots represent average values, their color indicating their distance to the extinction fit, where blue stands for values within one standard deviation and yellow for values outside of this range. The green lines at the bottom represent the boundaries of each of the shots taken by the camera. The shots have significant overlap in altitude.

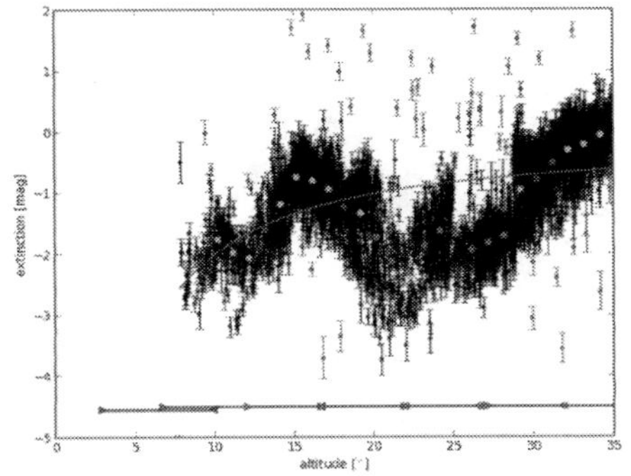

Figure 5. Example of an extinction plot for a cloudy event.

are summarized in Table 1. To compare the resulting rate of anomalous events to the Monte Carlo predictions, the trigger efficiency together with the duty cycle of the entire StS system need to be properly estimated. This procedure is not straightforward and is the subject of ongoing work within the Pierre Auger Collaboration.

Table 1. Number of StS events recorded by FRAM between March 2013 and April 2016.

complete events	no clouds	double bump trigger
1788	659	399

4 Conclusions

Finding a number of clearly anomalous events in a given energy bin could test the parameters of hadronic interaction models which are extrapolated from collider energies. The ratio of expected double-bump events depends both on the elasticity of the leading interactions and on the cross section of the deeply propagating particle. Some predictions using a current hadronic interaction model are presented in section 2. The number of false positives present in the dataset of the Pierre Auger Observatory dominates the expected number of actual anomalous events. A very efficient rejection of events affected by clouds is thus needed. A method making use of the FRAM telescope is described and the resulting number of candidate events is presented in section 3.2. Analysis of this dataset is ongoing. For example, another kind of false positive event may be created by fluctuations present in the electronics of the fluorescence detectors. This needs to be properly accounted for using a Monte Carlo approach. Moreover, the efficiency of the whole StS chain needs to be correctly estimated. Only then can the desired double-bump ratio be obtained.

The work is supported by the Grant Agency of the Czech Republic 14-17501S and by projects MSMT LG15014 and LM2015038.

References

[1] A. Aab *et al.* [Pierre Auger Collaboration], *Nucl. Instrum. Meth. A* **798**, 172 (2015), [arXiv:1502.01323 [astro-ph.IM]].
[2] K. H. Kampert and M. Unger, *Astropart. Phys.* **35**, 660 (2012).
[3] A. Aab *et al.* [Pierre Auger Collaboration], *Contributions to the 34th International Cosmic Ray Conference, (ICRC 2015)* [arXiv:1509.03732 [astro-ph.HE]].
[4] A. Aab *et al.* [Pierre Auger Collaboration], *Phys. Rev. D* **93**, no. 7, 072006 (2016) [arXiv:1604.00978 [astro-ph.HE]].
[5] *Proc. 15th ICRC 1977*, Plovdiv, 8 353, 1977.
[6] T. Bergmann *et al.*, *Astropart. Phys.* **26**, 420 (2007), [astro-ph/0606564].
[7] C. Baus, R. Engel, T. Pierog, R. Ulrich and M. Unger, *Proc. 32nd ICRC 2011*, [arXiv:1111.0504 [astro-ph.HE]].
[8] B. Keilhauer [Pierre Auger Collaboration], *Astrophys. Space Sci. Trans.* **6**, 27 (2010).
[9] J. Abraham *et al.* [Pierre Auger Collaboration]: *Astropart. Phys.* **33**, 108 (2010) [arXiv:1002.0366 [astro-ph.IM]].
[10] J. Ebr *et al.*, *Proceedings of the 3rd Astrorob workshop* **45**, 53 (2014).

Site Characterization of the Northern Site of the Cherenkov Telescope Array

Markus Gaug[1,2,⋆], *Lluís* Font[1,2], and *Camilla* Maggio[1,2]

[1] *Unitat de Física de les Radiacions, Departament de Física, Universitat Autònoma de Barcelona, Spain*
[2] *CERES-IEEC, Universitat Autònoma de Barcelona, Spain*

Abstract. We present the current knowledge and activities to assess the climatology of the "Observatorio del Roque de los Muchachos" (ORM), selected to host the Northern observatory of the CTA, with particular emphasis on molecular density profiles and aerosol extinction.

1 Introduction

Recently, the site selection process for the Cherenkov Telescope Array (CTA) [1] has been resolved. For its Southern part (CTA-S), a plateau of the Cerro Armazones in Northern Chile, close to Paranal was chosen[1], while the "Observatorio del Roque de los Muchachos" (ORM) on La Palma, Canary Islands, has been selected for its Northern observatory (CTA-N)[2].

The ORM looks also back on a history of more than 20 years of gamma-ray astronomy through Imaging Atmospheric Cherenkov Telescopes (IACTs), featuring first the HEGRA[3] and then the MAGIC Telescopes[4]. The latter dedicated considerable efforts to characterize the atmosphere above with respect to molecular and aerosol profiles, and to clouds [2–4]. Moreover, existing optical telescopes are active in this task [5, 6], and the very same host organization of the ORM, the "Instituto de Astrofísica de Canarias" (IAC), through their "Sky Quality Group" (SQG)[5] [7–9]. In addition, also the "European Extremely Large Telescope" (E-ELT) have carried out site testing and characterization campaigns on La Palma [10]. Moreover, the only about 150 km apart "Observatorio del Teide" (OT)[6], and particularly the "Izaña Atmospheric Research Center" (IARC)[7], or simply Izaña Observatory (IZO), both located on the neighboring island of Tenerife at about 2400 m a.s.l., have monitored atmospheric conditions over large time series. Atmospheric conditions between Tenerife and La Palma, at these altitudes, are highly correlated [11], and data from the IARC can hence be used, at least to better understand the aerosol characteristics, at the ORM.

The atmosphere in the sub-tropical region of the Canary Islands is characterized by great stability throughout all the year. This is due to the combination of two processes of large-scale atmospheric circulation [12]: the descending branch of the Hadley cell around 30° N and the "Trade" or "Alisios" winds, coming from the Azores high area, that blow in the low troposphere, above an ocean which is relatively cold. As a result, a temperature inversion layer (TIL) appears above 1300 m a.s.l. in Winter (~1000 m a.s.l. in Summer)[8] [13], which can be usually well identified by the "sea of stratocumulus" at its base. This layer separates two well-defined regimes: below it, the moist marine boundary layer (MBL) and above it, the dry free troposphere (FT). The TIL is present in > 84% of the time in the Western Canary Islands (> 90% observed at Guimar, Tenerife; the closest sounding station to La Palma), and shows moderate seasonal variation, with more frequent break-downs observed in February and November, probably modulated by the subsidence of the Hadley cell [13].

The ORM is located at an altitude between 2000 m and 2400 m a.s.l., hence well above the thermal inversion layer. This ensures that the installations are in the FT. At the ORM, clean air and clear sky conditions are hence prevailing all around the year. The reason is, by the way, twofold: apart from being located above the stable inversion layer, it is situated on an island far away from any significant industrial activity.

Now that site selection is accomplished and expected average observational duty cycle and performance established [14], the CTA-N is primarily interested in assessing molecular density profiles up to about 25 km a.s.l. and their variations, in understanding aerosol profiles and their optical characteristics, in order to trim its atmospheric calibration strategy [15].

⋆e-mail: markus.gaug@uab.cat

[1] see www.cta-observatory.org/about/array-locations/chile/
[2] see www.cta-observatory.org/about/array-locations/la-palma/
[3] see www.mpi-hd.mpg.de/hfm/CT/CT.html
[4] see magic.mpp.mpg.de/
[5] see www.iac.es/eno.php?op1=4&lang=en and www.iac.es/proyecto/site-testing/
[6] see www.iac.es/eno.php?op1=3
[7] see izana.aemet.es/index.php?lang=en
[8] About 25% of the time, additional TIL's are present above the primary one mentioned here [13].

DOI: 10.1051/epjconf/201714401010

2 Molecular density profiles

We studied 3 years (2013–2015) of now-cast predictions of the "Global Data Assimilation System" (GDAS)[9], available in time steps of 3 hours UTC for La Palma, more specifically for the grid point 29°N, 18°W, about 28 km northwards the ORM, above the Atlantic, smoothly joint with the NRLMSISE-00 data[10], for lat. 28.77°, lon. 342.11° above 20 km a.s.l. A spatially better resolved WRF model [16], as employed by J. Marín in these proceedings [17], has not yet been tried, however the European ECMWF[11] is currently being tested, yielding similar or even better results, while the IG2 model [19] does not agree well with instantaneous MAGIC data at these altitude ranges, at least what concerns the temperature profiles. Similarly, the NRLMSISE-00 model [20], although very accurate above 20 km a.s.l., is not able to resolve all tropospheric temporal variations and can show deviations from the GDAS model of up to 10%.

The GDAS data set is available in 24 pressure levels[12], each associated with an individual geopotential height, a temperature and other data, such as wind components and humidity. Geopotential heights have been converted to local altitudes a.s.l. using the WGS-84 ellipsoid for the evaluation of the combined effects of gravitational and centrifugal forces on the local gravity at a given altitude and latitude, following the prescription of Mahoney [21]. Values between these levels were interpolated. One has to keep in mind however, that fine-structures possibly present in the lower troposphere, and particularly in the tropopause (see e.g. Fig. 1 of [13] or Fig. 1 of [22]), cannot be reproduced by such an interpolated coarse grid.

For the purpose of ground validation, temperature and pressure values measured by the MAGIC weather station were compared with the predicted ones at the altitude of the weather station, which was assessed by a specialized company using a GPS survey in 2015. Fig. 1 shows all entries obtained for night-times. For the pressure comparison a bias of (-0.25 ± 0.01) mbar was obtained, compatible with no bias, once the accuracy of the MAGIC pressure sensor (of ± 0.8 mbar) is taken into account. The temperature comparison yields a significant bias of (2.08 ± 0.02)°C, which can however be understood by ground cooling effects, that are not present at the chosen GDAS grid point (lying above the Sea). Using only day-time data, an opposite effect is observed. Typically such ground effects are limited to low altitudes. Nevertheless, even in the exaggerated assumption of being constant throughout the troposphere, this temperature bias would alter the density profile by only 0.8%. Another direct validation was performed with the MAGIC LIDAR data, which shows pure molecular back-scattering and extinction signatures throughout most of the probed altitude ranges, and where contemporaneous GDAS profiles yield accurate χ^2-fit distributions, in contrast to average molecular profiles [23].

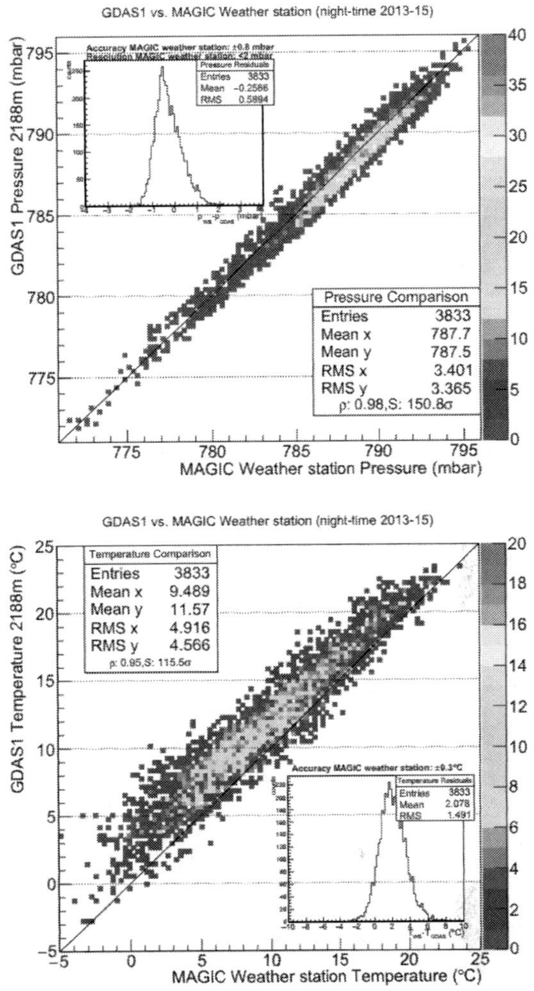

Figure 1. Cross-correlation of the GDAS predicted pressures and temperatures at 21h, 0h, 3h and 6h UTC at the altitude of MAGIC, with the simultaneous MAGIC weather station data. The inlets display the distribution of the respective differences.

Fig. 2 shows averaged density profiles, grouped according to the findings of [22]: an "average winter" (DEC,JAN,FEB,MAR,APR), characterized by multiple thermal tropopauses, which are several kilometers thick and can reach extremely high altitudes of up to 20 km a.s.l.. This typical mid-latitude tropopause is influenced by the sub-tropical jet stream (STJ) centered at lower latitudes. The "average summer" (JUL,AUG,SEP) model is characterized by only one thermal tropopause, typical for a tropical upper troposphere, with the STJ having moved northwards of the Canary Islands. The intermediate months (MAY,JUN,OCT,NOV) show transition behavior between both scenarios. Tropical tropopauses occur nevertheless also occasionally during the rest of the year. The lower end of the multi-layered, thick mid-latitude tropopauses move slightly upwards during spring and de-

[9]https://www.ncdc.noaa.gov/data-access/model-data/model-datasets/global-data-assimilation-system-gdas
[10]ccmc.gsfc.nasa.gov/modelweb/atmos/nrlmsise00.html
[11]http://weather.unisys.com/ecmwf/index.php
[12]1000, 975, 950, 925, 900,850, 800, 750, 700, 650, 600, 550, 500, 450, 400, 350, 300, 250, 200, 150, 100, 50, 20 mbar, and the surface level.

scend again in autumn, while the single-layered tropical tropopauses show much less seasonality [22]. From these average scenarios, and the extremes also seen in Fig. 2, the input models for detailed simulations have been created [24].

The most significant differences between both averages are found below 5 km a.s.l. and in the tropopause. The latter has also been observed in the South [17]. At about 15 km, the strongest relative differences between Summer and Winter is observed, with an average seasonal variation amplitude of about 7% (see Fig. 2 bottom). Diurnal variations are more than an order of magnitude smaller at all altitudes above CTA-N.

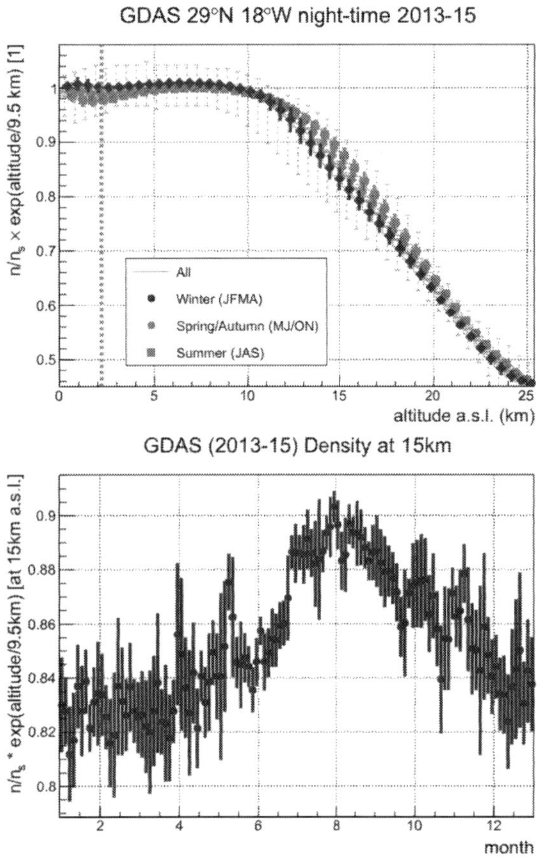

Figure 2. Top: density profiles of 3-years GDAS nowcasts at 21h, 0h, 3h and 6h UTC, scaled to the U.S. Standard Atmosphere density $n_s = 2.55 \cdot 10^{19}$ cm^{-3} and multiplied with the exponential of an average scale height ($H_s = 9.5$ km) for better visibility. The dark blue, red and green points display an "average Winter", "average Spring/Autumn" and "average Summer" scenario, respectively. Error bars represent the standard deviation, while the light blue lines show the peak-to-peak differences for the entire data set. Bottom: seasonal variation of density evaluated at 15 km a.s.l.

Ozone intrusions into the upper troposphere, the "stratosphere-troposphere exchange" (STE) processes, have not been found for the Western Canary Islands, as assessed climatologies of the "Potential Vorticity" (PV) parameter [18]. Nevertheless, a 20 yr ozone-sonde campaign revealed that the extra-tropical tropopause above Tenerife shows increased ozone concentration during late Winter and Spring in the range from 12 to 15 km a.s.l. [22], well below the stratospheric ozone layer above ~18 km.

3 Aerosol Profiles

Aerosols have been characterized at the ORM with the help of elastic LIDARs [10, 23], dust counters [6] and extinction measurements by optical telescopes [8]. Moreover, several types of particulate analyzers and nephelometers are employed at the Izaña station on Tenerife [25–27] and on airplanes [28]. Sun photometer inversion products are available for Izaña through the AERONET network[13]. Also satellite information has been investigated [11] and back-trajectories simulated, using the HYSPLIT code [10, 29].

Ground-layer aerosol at the Western Canary Islands, of marine/oceanic origin with small admixture of anthropogenic aerosol [10] typically shows a bi-model size distribution during clear nights, with a size distribution peaking at 50 nm during night (see Fig. 7 of [25]), an exponential decay of density, with scale heights around 500 m [23] and aerosol optical thickness (AOT) of ~0.02 at 532 nm [10, 23]. In this situation, only Aitken and Accumulation modes of aerosol are found while the Coarse mode is practically absent. This leads to typical Ångström exponents of 1.5–2.0 [26, 30] and a single scattering albedo of 0.91 [27] and LIDAR ratios around 30 sr [31]. These findings are compatible with the median value for the V-band extinction of 0.13 mag airmass^{-1} found by [8], translating into an AOT of about 0.02.

The situation changes drastically during dust intrusion events from the Western Sahara (the so-called "calima"), when mean particle sizes reach 1 μm or greater [6]. Such events occur during about 30% of the Summer nights [8], and rarely during the rest of the year. Typical LIDAR ratios are then ~60 sr [31], and AOTs can reach values in excess of 0.5 [6]. The calima-related dust can reach altitudes of and higher than 7 km a.s.l. [10], although lower altitudes are more frequent [23].

Fig. 3 shows the probability to observe a given value of AOT from different altitudes above the CTA-N, obtained with a 2-years statistics of the MAGIC LIDAR.

4 Conclusions

The atmosphere above the ORM has been characterized by a large variety of instruments in the past. For molecular and aerosol profiles, detailed climatological studies exist providing a deep understanding of their origins and characteristics. The CTA-N profits from these to design its atmospheric calibration strategy and reduce systematic uncertainties to hitherto unprecedented levels.

[13]http://aeronet.gsfc.nasa.gov/cgi-bin/type_one_station_opera_v2_inv2?site=Izana&nachal=0&level=2&year=-1&month=0&day=1&aero_water=0&if_day=0&if_err=0&year_or_month=0&place_code=0

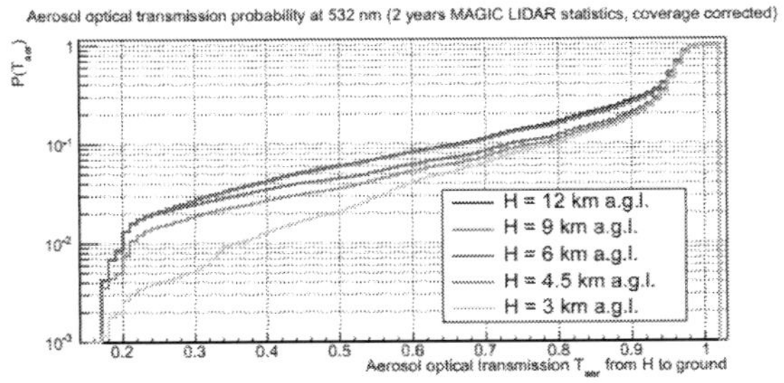

Figure 3. Aerosol Transmission probability for different emission altitude above the MAGIC LIDAR. Differences between the yellow line (emission height 3 km a.g.l.) and the red line (emission height 6 km a.g.l.) are mainly due to calima, whereas clouds are frequently found between 4.5 km and 9 km a.g.l.

References

[1] M. Actis et al., *Design concepts for the Cherenkov Telescope Array CTA: an advanced facility for ground-based high-energy gamma-ray astronomy*, Exp. Astron. **32**, 193 (2011).

[2] C. Fruck et al., *A novel LIDAR-based Atmospheric Calibration Method for Improving the Data Analysis of MAGIC*, Proc. 33rd ICRC **1054** (2013), arXiv:1403.3591

[3] M. Gaug et al., *Atmospheric Monitoring for the MAGIC Telescopes*. Proc. 1st AtmoHEAD Conf. (2014).

[4] C. Fruck & M. Gaug, *Atmospheric monitoring in MAGIC and data corrections*, Proc. 2nd AtmoHEAD Conf., EPJ **89** 02003 (2015).

[5] G. Lombardi et al., *El Roque de Los Muchachos Site Characteristics. II. Analysis of Wind, Relative Humidity, and Air Pressure*. PASP **119** 292 (2007).

[6] G. Lombardi et al., *El Roque de Los Muchachos site characteristics. III. Analysis of atmospheric dust and aerosol extinction* A&A **483** 651 (2008).

[7] C. Muñoz-Tuñón et al., *Night-time image quality at Roque de los Muchachos Observatory*. Astron. Astrop. Suppl. Ser. **125** (1) 183 (1997).

[8] A. García-Gil et al., *Atmosphere Extinction at the ORM on La Palma: A 20 yr Statistical Database Gathered at the Carlsberg Meridian Telescope*. PASP **122** 1109 (2010).

[9] B. García-Lorenzo et al. *Infrared astronomical characteristics of the Roque de los Muchachos Observatory: precipitable water vapour statistics* MNRAS **405** (4) 2683 (2010).

[10] M. Sicard et al., *Results of site testing using an aerosol, backscatter lidar at the Roque de los Muchachos Observatory* MNRAS **405** (1) 129 (2010).

[11] A. M. Varela et al., *Astronomical site selection: on the use of satellite data for aerosol content monitoring* MNRAS **391** (2) 507 (2008).

[12] E. Palmen & C. W. Newton. *Atmospheric Circulation Systems - Their Structure and Physical Interpretation* Intern. Geophys. Ser. **13** Acad. Press (1969).

[13] J. Carrillo et al., *Characterization of the marine boundary layer and the trade-wind inversion over the sub-tropical North Atlantic* Boundary-Layer Meteor. **158** (2) 311 (2016).

[14] T. Hassan et al., *Monte Carlo Performance Studies of Candidate Sites for the Cherenkov Telescope Array*, subm. to Astrop. Phys. (2017).

[15] M. Gaug, *CTA Atmospheric Calibration*, these proc.

[16] W. C. Skamarock et al., *A Description of the Advanced Research WRF* NCAR/TN-468+STR (2008).

[17] J. Marín et al., *Density profile characterization and modeling at Paranal and Armazones 2k sites*, these proc.

[18] E. Cuevas et al., *Surface O3 characterization in the sub- tropical North Atlantic troposphere*, Atmos. Chem. Phys. **13** 1973 (2013).

[19] J. J. Remedios et al., *MIPAS reference atmospheres and comparisons to V4.61/V4.62 MIPAS level 2 geophysical data sets* Atm. Chem. Phys. Disc. *7* 9973 (2007).

[20] J. M. Picone et al., *NRLMSISE-00 empirical model of the atmosphere: Statistical comparisons and scientific issues*. J. Geophys. Res.: Space Phys. *107* (A12) SIA 15 1 (2002).

[21] M. J. Mahoney. *A discussion of various measures of altitude*, NASA Jet Propulsion Laboratory, http://mtp.mjmahoney.net/www/notes/altitude/altitude.html (2001).

[22] J. J. Rodriguez-Franco & E. Cuevas, *Characteristics of the subtropical tropopause region based on long-term highly resolved sonde records over Tenerife*. J. Geophys. Res.: Atmospheres **118** (19) 10754 (2013).

[23] C. Fruck et al., *Characterizing the atmosphere over the MAGIC site, analyzing LIDAR data with a novel signal inversion algorithm*, in preparation.

[24] M. Vrastil et al., *Overview of Atmospheric Simulation Efforts in CTA*, these proc..

[25] S. Rodríguez et al., *Atmospheric nanoparticle observations in the low free troposphere during upward orographic flows at Izaña Mountain Observatory* Atmos. Phys. Chem. **9** 6319 (2009).

[26] H. Maring et al., *Aerosol physical and optical properties and their relationship to aerosol composition in the free troposphere at Izaña, Tenerife, Canary Islands, during July 1995*, J. Geophys. Res. **105** 14677 (2000).

[27] E. Andrews et al. *Climatology of aerosol radiative properties in the free troposphere* Atmosph. Res. **102** 365 (2011).

[28] D. R. Collins et al. *In situ aerosol-size distributions and clear-column radiative closure during ACE-2* Tellus **52B** 498 (2000).

[29] S. Rodríguez et al. *Transport of desert dust mixed with North African industrial pollutants in the subtropical Saharan Air Layer* Atmos. Chem. Phys. Disc. **11** 8841 (2011).

[30] D. Whittet et al., *The extinction properties of Sahara dust over La Palma* Vistas in Astronomy **30** 135 (1987).

[31] A. Ansmann & D. Müller, *LIDAR Range-Resolved Optical Remote Sensing of the Atmosphere*, p. 112, Springer, Berlin (2005).

Aerosol Measurements with the FRAM Telescope

Jan Ebr[1,⋆], *Jiří* Blažek[1], *Michael* Prouza[1], *Dušan* Mandát[1], *Miroslav* Pech[1] for the Pierre Auger Collaboration[⋆⋆], *Martin* Mašek[1], *Jakub* Juryšek[1], *Petr* Janeček[1], *Petr* Kubánek[1], *Jiří* Eliášek[1,4], *Martin* Jelínek[2], and *Ivana* Ebrová[3]

[1] *Institute of Physics of the Czech Academy of Sciences, Prague*
[2] *Astronomical Institute of the Czech Academy of Sciences, Ondřejov*
[3] *Nicolaus Copernicus Astronomical Center, Polish Academy of Sciences, Warsaw*
[4] *Faculty of Mathematics and Physics, Charles University, Prague*

Abstract. Precision stellar photometry using a telescope equipped with a CCD camera is an obvious way to measure the total aerosol content of the atmosphere as the apparent brightness of every star is affected by scattering. Achieving high precision in the vertical aerosol optical depth (at the level of 0.01) presents a series of interesting challenges. Using 3.5 years of data taken by the FRAM instrument at the Pierre Auger Observatory, we have developed a set of methods and tools to overcome most of these challenges. We use a wide-field camera and measure stars over a large span in airmass to eliminate the need for absolute calibration of the instrument. The main issues for data processing include camera calibration, source identification in curved field, catalog deficiencies, automated aperture photometry in rich fields with lens distortion and corrections for star color. In the next step, we model the airmass-dependence of the extinction and subtract the Rayleigh component of scattering, using laboratory measurements of spectral sensitivity of the device. In this contribution, we focus on the caveats and solutions found during the development of the methods, as well as several issues yet to be solved. Finally, future outlooks, such as the possibility for precision measurements of wavelength dependence of the extinction are discussed.

1 Motivation

The Pierre Auger Observatory [1] is the world's largest detector of ultra-high energy cosmic rays (UHECR). The extensive air showers (EAS) initiated by the UHECR are observed using an array of surface detectors, which register the secondary particles that reach the ground, and 27 optical telescopes which, during sufficiently clear and dark nights, observe the fluorescence light emitted during the passage of the EAS through the atmosphere. The later method is particularly sensitive to the transmission properties of the atmosphere, as showers are observed to a distance of tens of kilometers from the telescopes. To monitor those properties, many devices, such as cloud cameras, elastic and Raman LIDARs, calibration lasers, all-sky cameras and the FRAM telescope are installed on the site. The FRAM telescope consists of narrow-field (23×15 arcmin^2) and wide-field ($7^\circ\times7^\circ$) optical setups comounted on a commercial astronomical mount and operated as a fully autonomous robotic observatory. Currently the wide-field camera is mostly used for atmospheric monitoring [2].

The main purpose of the FRAM telescope in its current setup is part of the so-called rapid monitoring program [3], which focuses on targeted measurements of atmospheric properties triggered by the observation of specific EAS in the fluorescence telescopes (this method is often called Shoot-the-Shower, or StS, by the Auger Collaboration). The main goal of this program is the observation of showers with anomalous longitudinal light profiles – for each candidate for such shower, it must be first determined whether the measured fluorescence profile is affected by clouds or other inhomogeneities in the atmosphere. For this purpose, the FRAM telescope has an advantage over the laser-based measurements, because it does not emit any light and thus does not interfere with the operation of the fluorescence telescopes. When a trigger is received for an interesting shower, the wide-field camera of the FRAM telescope is used to measure the extinction of starlight along the observed path of the shower on the sky, taking several (depending on the orientation of the shower, but at least six) images across the whole field of the fluorescence telescopes, that is between 30 and 2 degrees in altitude above the horizon [4]. In principle, the difference between measured brightness of a star and that from a catalog can be written as

$$m_{\text{inst}} - m_{\text{cat}} = Z + kA + f(...) \qquad (1)$$

where Z is the zeropoint (a calibration constant of the system), A is airmass (amount of air traversed by the light relative to a star in zenith), k is the extinction coefficient ($1.086\times$ the vertical optical depth) and $f(...)$ is in general a correction function of some properties of the star and the optical system. Local deviations from this formula are

⋆e-mail: ebr@fzu.cz
⋆⋆Full author list: http://www.auger.org/archive/authors_2016_09.html

an indication of the presence of clouds or inhomogeneous aerosol layers.

A curious reader could ask, what are the values of the extinction coefficient and whether this method could be used to determine this value and thus the vertical aerosol optical depth (VAOD). The answer is a qualified yes – while extracting the value from the fit is easy, reaching a precision good enough that the data become relevant for atmospheric calibration requires significant effort. In this paper, we describe this effort and the current status of the method, focusing on aerosol measurements. Two main things should be kept in mind: firstly that this is still a work in progress and secondly that the method has specific limitations, in particular in the fact that we always measure the integral extinction across the whole atmosphere and cannot determine the height profile of the aerosols.

2 Photometry and extinction

First we introduce and clarify some quantities. The light of a star reaches the Earth as an apparent flux J in Wm^{-2} or similar units. The apparent brightness m is defined in logarithmic units – magnitudes – with respect to a reference flux by

$$m = -2.5\log_{10}(J/J_{\mathrm{ref}}). \quad (2)$$

The physical flux is detected by a detector, in our case a CCD chip, where it is registered as a current and digitized, resulting in an instrumental flux J_{inst} in arbitrary units. The CCD camera is a highly linear detector (after data reduction) and thus

$$J = CJ_{\mathrm{inst}} \quad (3)$$

for some calibration constant C. For apparent brightness, we get

$$m = m_{\mathrm{inst}} - 2.5\log_{10}(C/J_{\mathrm{ref}}) = m_{\mathrm{inst}} - Z \quad (4)$$

where Z is called the zeropoint and is the primary calibration constant of any photometric device.

In the atmosphere, light suffers exponential extinction

$$J_{\mathrm{ground}} = J_{\mathrm{inst}}\exp(-\tau) \quad (5)$$

where the optical depth τ can be easily translated into a difference in apparent brightness because

$$\tau = \ln(J_{\mathrm{space}}/J_{\mathrm{ground}}) = 0.921\Delta m. \quad (6)$$

For a given direction, the optical depth can be related to the vertical optical depth (VOD) by

$$\tau = \tau_{\mathrm{vertical}}A \quad (7)$$

where the dimensionless quantity A is called airmass and quantifies the relative amount of air traversed by the light with respect to a star at the zenith. For a flat Earth it would have a 1/cos dependence on zenith angle θ, for curved Earth the following approximation is valid up to 89°:

$$A = \frac{1}{\cos\theta + 0.025\exp(-11\cos\theta)}. \quad (8)$$

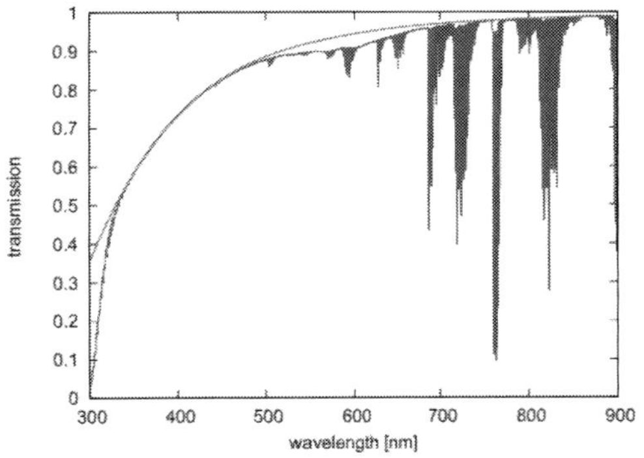

Figure 1. Total vertical transmission of aerosol-free atmosphere at the Auger site calculated in MODTRAN, courtesy of Bianca Maria Dinelli and Enzo Papandrea, Istituto di Scienza dell'Atmosfera e del Clima, Bologna. Green: Rayleigh scattering only. Magenta: Rayleigh scattering and molecular absorption.

We further define the extinction coefficient k so that it corresponds to the change of apparent brightness by one VOD

$$k = \frac{\tau_{\mathrm{vertical}}}{0.921} = 1.086\tau_{\mathrm{vertical}}. \quad (9)$$

For a given star, we thus expect the observed magnitude to be related to the catalog one as

$$m_{\mathrm{inst}} = m_{\mathrm{cat}} + Z + kA. \quad (10)$$

As the optical depth and extinction coefficient are numerically very close, we use them somewhat interchangeably in the text, however taking care to include the correct factor in actual calculations. In reality, the extinction has several components. In the blue visible light where FRAM mostly operates, the dominant contribution is the Rayleigh scattering off air molecules, while the molecular absorption is negligible (Fig. 1). When the Rayleigh contribution is subtracted from the VOD, the rest is thus assumed to be due to aerosols and is referred to as the vertical aerosol optical depth, VAOD.

In principle, we could extract the extinction coefficient from a single observation of a single star, providing that we know the zeropoint precisely enough. This was the idea of the photoelectric photometer installed on FRAM until 2010, but the technical problems related to its robotic use proved insurmountable. For CCD photometry, such approach is not feasible because of the variations in the zeropoint due to many effects, if one wants to achieve precision higher than of the order 0.1 in optical depth, which is quite coarse considering that 0.1 is the usual VAOD cut in analyses at the Auger Observatory. Therefore, our goal must be to measure the VAOD at the level of at least 0.01.

To do so requires to measure $m_{\mathrm{inst}} - m_{\mathrm{cat}}$ as a function of A and extract both k and Z simultaneously. We first tried to follow some reference stars during the night as they naturally pass through a range of airmasses, but we found out

DOI: 10.1051/epjconf/201714401011

that probably both k and Z vary considerably during the time needed. Moreover this approach makes it difficult to deal with systematic effects caused by star color and its inevitably changing position on the CCD chip. This had led us to the wide-field approach, where we measure a lot of stars in different airmasses within a short time window and we deal with systematics and outliers within a complex fitting procedure.

3 Data selection

As we already mentioned, the motivation for the aerosol measurements came from the rapid monitoring program. In this program, we take pictures along the observed sky path of an interesting shower. To fully control the atmospheric influence, we focus not only on the directions where the shower was observed in the fluoresence telescopes, but scan the whole path of the shower across their field of view to make sure that the non-detection of the shower in some parts is not due to a cloud. Thus every shower triggers a scan in altitude covering a swath of sky between 2 and 30 degrees above the horizon, which is potentially a huge span of airmasses – the airmass at 30 degrees altitude is 2 whereas it is roughly 40 at horizon. The useful span is however limited by our ability to detect stars dimmed by very high extinction and also the fact that near to the horizon we are looking to very large distances, increasing the probability to encounter a distant cloud. Such altitude scans provide a quite homogeneous dataset for aerosol measurements but are limited to periods when the rapid monitoring program was working (and interesting showers were detected). Since the beginning of 2016, we started to take additional altitude scans at frequent interval during all nights (even during full Moon when the fluorescence telescopes do not operate at all).

Only those scans that do not show any cloud contamination at all (apart from very low altitudes) are useful for the aerosol analysis, because the presence of clouds makes it very difficult to fit the airmass dependence of extinction. The present work is based on scans taken between March 2013 and April 2016, from which 657 clear scans were selected. These were picked by hand using the plots generated for the rapid monitoring program and then further cleared taking into account the quality of fit, removing any suspicious artifacts and making sure that the camera temperature was properly set for all the images. Since we started the dedicated observations at the beginning of 2016, there is a wealth of new data (likely at least doubling the dataset) awaiting analysis. All the data shown here are based on images taken in the B filter with 30-second exposures with the wide-field camera of the FRAM telescope, which is a Moravian Instruments G4-16000 CCD with a $36mm \times 36$ mm chip mounted on a Nikkor 300/2.8 lens. This setup provides a rectangular field of view $7° \times 7°$.

4 Data reduction

The CCD chip converts incoming photons to electrons and those are then read out for each pixel and the signal is amplified and digitized in an A/D converter. Thus the signal

Figure 2. An example of a photo in a rich star field, not corrected for flat field, shows the significant vigneting in corners that must be removed using a flat field image or by a fit to the stellar data.

for each pixel (in ADU) is in principle linearly dependent on the incoming light into the pixel. In reality, the signal is a combination of the light of the star, the night-sky background and other scattered light, the dark current of the chip, and the noise of the electronics. The electronics contribution is very small for Moravian Instruments cameras and also the dark current (the signal recorder with zero light, mostly the thermal current of the semiconducting chip and thus dependent on temperature) is also very small for most of the pixels when cooled to $-20°$, with the exception of typically a thousand of "hot" pixels that can even become saturated by the thermal current when the exposure is sufficiently long. The dark current can be dealt with by taking a series of exposures with the shutter closed ("dark frames") and subtracting the average or median of those exposures from the light images, or, if it is sufficiently low, by detecting the hot pixels in the light images and removing them.

Most of the background light is removed by the aperture photometry method (see later), but the dark-current correction plays an important role when we take into account that the sensitivity of the lens-camera system is not uniform across the field of view, because of the different sensitivity of each pixel (a small effect), the vigneting of the lens (a large effect, up to 50 %, Fig. 2) and due to dust and other contamination on various optical surfaces. Normally this is dealt with using the "flat field", an image of a uniformly lit surface. Thus for a CCD image the following holds schematically

$$image = flat \times original + dark \tag{11}$$

and thus one would like to retrieve the original image from the one taken by

$$original = \frac{image - dark}{flat}. \tag{12}$$

DOI: 10.1051/epjconf/201714401011

Figure 3. Aperture photometry: the inner circle is used to calculate the light flux and the annulus defines the background.

We observe that the dark current slowly increases during the night, even when the temperature is reported to be constant (probably due to the fact that the sensor is not directly on the chip and the chip heats up when used). A single dark image is noisy and thus many have to be taken to create a master dark image used in calibration. This is normally done in the evening or in the morning, but such master dark image either undercompensates or overcompensates the dark current and that affects the flat subtraction. While this is a small effect, we are currently developing methods to deal with it, such as interpolation between evening and morning dark frames. The flat field correction is never perfect anyway, because for a wide-field camera, the evenly lit surface is hard to realize and thus we use the dawn and dusk sky, which has its own drawbacks. We correct for the flat-field deficiencies that are mostly radial, using the data.

After the dark and flat corrections, the flux for each star is evaluated using aperture photometry. In this method, a circle of a fixed size is centered on each detected light source in the image and the light in all the pixels is summed. Another larger circle is then taken concentrically and the signal in the annulus is used to estimate the background to be subtracted from the previous value (Fig. 3). The aperture needs to be large enough so that most of the point spread function is contained in the aperture – taking an aperture too small leads to nonlinearity of the measurement, because for bright stars the amount of light lost outside the aperture becomes more significant. On the other hand, taking the aperture too large means a danger of contamination with the light of another nearby star. We avoid this problem by rejecting any pair of detected sources that is closer together than twice the aperture, but this leads to significant losses of observable stars in some of in star-rich areas of the Milky Way.

Another interesting issue with aperture photometry arises when the shapes of the images of stars are not constant across the field of view. The coma and astigmatism that appear for stars farther from the center of the image are actually not a big issue, because their dependence on the position on the chip is radial and thus mimic the effect of the vignetting which we correct for using the data anyway. A bigger problem was found by visually inspecting the images, when it became apparent that the stars have the tendency to be less sharp at one side of the image. We attribute this behavior to mechanical flexing between the lens and the camera and have since improved the mechanical support accordingly. We have found that the increase in the FWHM of the stars from one side of the image to the other correlates with the decrease of their observed brightness (Fig. 4). This effect cannot be corrected for in the data because it can appear in the same direction as does the atmospheric extinction that we are trying to fit.

5 Fitting stars with a model

When the light flux from the stars has been measured, we need to identify those in the catalog. The wide field of view of the camera is distorted and thus a simple linear identification algorithm fails. While tools to identify stars in curved fields in principle exist, we were not able to make them work properly for our data. Instead of writing our own method, we have resorted to a simpler solution: we cut the images into smaller sub-images and within those the distortion is small enough so that the linear identification works. We have tested various configuration, monitoring the distances between the measured and catalog sources and we have concluded that cutting the image based on a 4×4 grid of equally sized tiles and merging the middle 2×2 into one works very well (Fig. 5). This method is slightly problematic in cloudy weather, when some of the tiles may not have enough stars to properly identify the star fields, but that is not relevant to the aerosol analysis.

The comparison with the catalog is further complicated by different star colors. In astronomy, it is customary to measure star brightness in one of pre-defined "filters", which nowadays mean a complete specification of a bandpass. The most widespread set of filters is the Johnson UBVRI system, to which FRAM with its physical color filters is roughly compliant. However the Tycho2 catalog that we use (as the only homogeneous all-sky catalog for the relevant range of star magnitudes available at the time when we started the analysis) uses B and V filters that are significantly different from the Johnson standard. We fit the dependence of the brightness observed by FRAM on the catalog values given by Tycho using our data and we have found that

$$B_{\mathrm{FRAM}} \approx 0.7 B_{\mathrm{Tycho}} + 0.3 V_{\mathrm{Tycho}} \tag{13}$$

where this formula holds only for stars with $B - V < 0.8$, because at higher values of $B - V$ (more red stars) another population appears. This is likely a real physical effect of star populations in the Galaxy. We are currently investigating the option to use a new all-sky photometric catalog APASS from AAVSO which may make the color calibration easier and also allow us to extend the measurements to the R filter where the Tycho2 catalog has no data. We will come to the problems caused by the stellar spectra later.

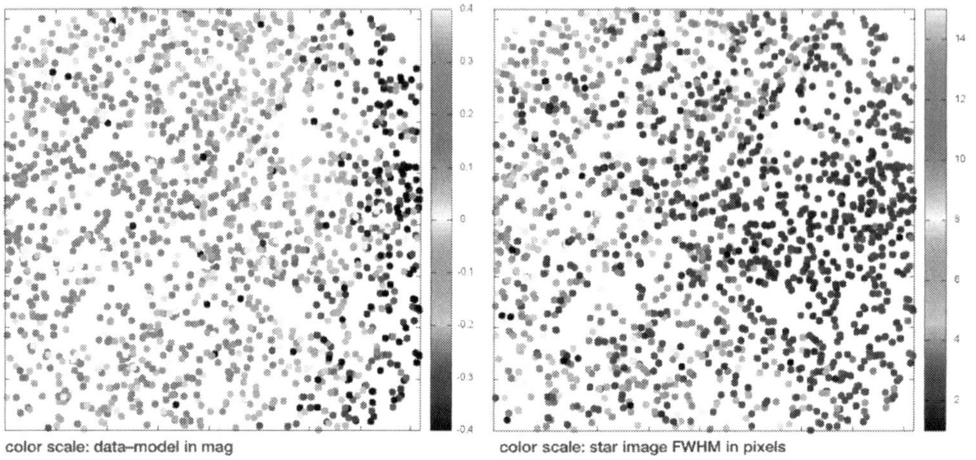

Figure 4. The effect of the varied star FWHM caused by mechanical problems between camera and lens (right) on the deviation of the measured brightness of those stars from the model (left).

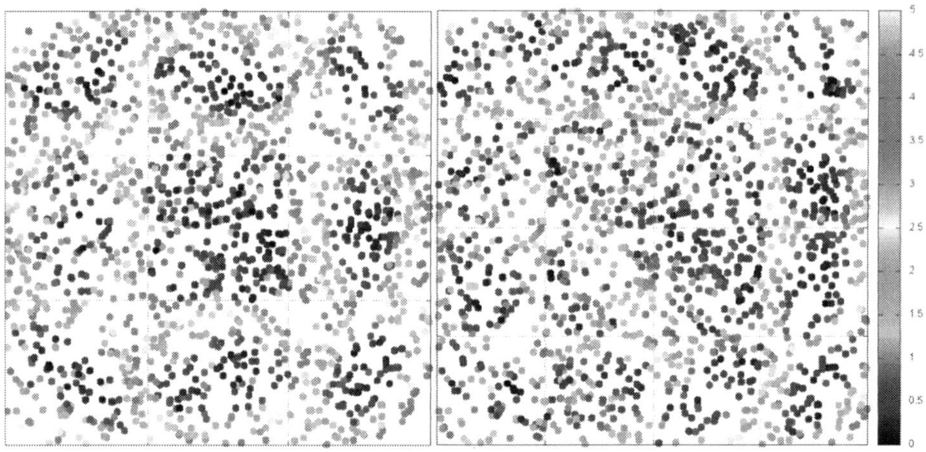

Figure 5. Distances between detected sources and catalog positions of stars in pixels (color coded) when the source identification was done after splitting into 3 × 3 tiles (left) and 4 × 4 tiles with inner 2 × 2 merged (right).

Putting all the effects we have mentioned together, we can formulate the following model for the observed brightness of a star:

$$m_{\text{inst}} = m_{\text{cat}} + Z_i + k_i A + c_1(B-V)(c_2(B-V)+1) + R_1 r(R_2 r + 1) + k_C A(B-V) \quad (14)$$

where B and V are magnitudes of the star in the catalog in the respective filters, r is the distance of the star from the center of the respective image and M, c_1, c_2, R_1, R_2 are parameters of the system (optics+CCD+data reduction). The last term represents the dependence of the extinction on the star color (and we will revisit it later).

We fit this model to many scans at once. For each scan (which comprises several images) we allow a different pair of Z and k (thus the index i) while the other parameters are the same for all scans – even k_C, because we expect the color-dependence to be dominated by the Rayleigh scattering, which is constant. We take only stars brighter than 9.5 mag because of the unreliability of the Tycho2 catalog below this value and only stars fainter than 6.5 mag because of saturation. The time needed for the multi-dimensional fit increases with the number of parameters; we have found that we can comfortably fit about 200 scans at once. We perform this procedure in two iterations, cutting obvious outliers after the first iteration. We have tested that the multi-dimensional fitting algorithm is reliable by fitting

DOI: 10.1051/epjconf/201714401011

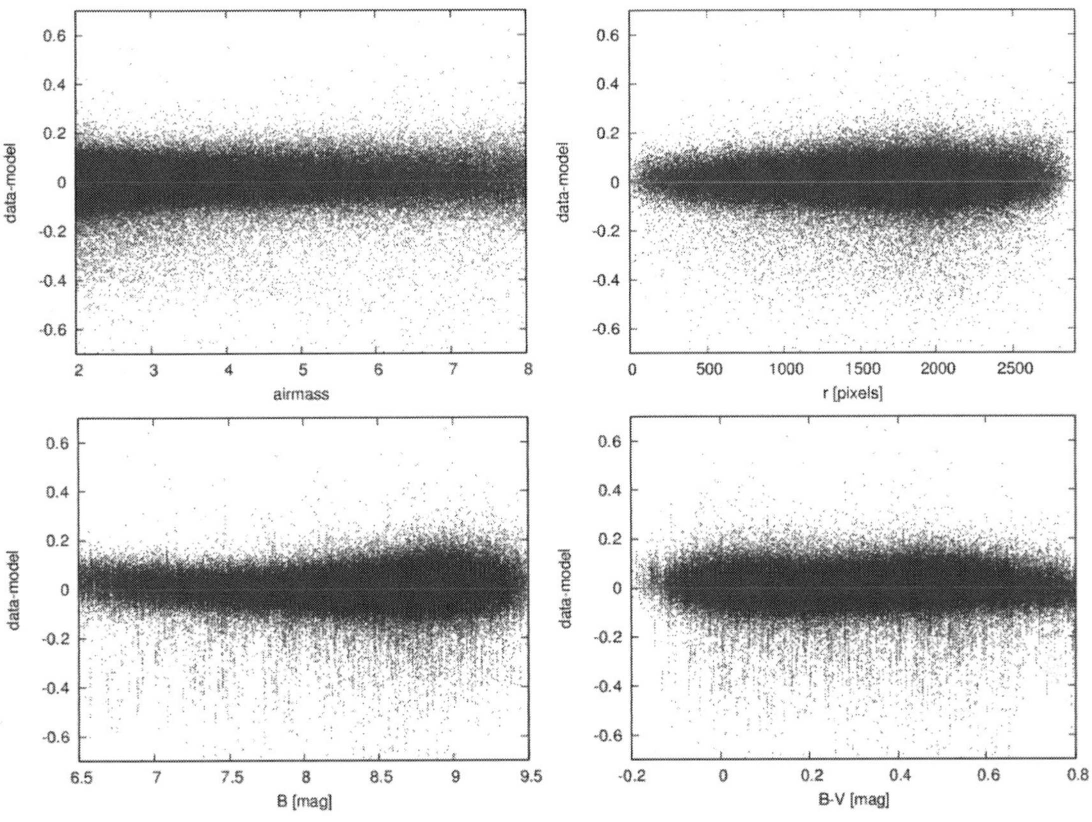

Figure 6. Residua of the global fit as a function of various variables.

only Z and k shower by shower using the system parameters from the global fit. The parameter M comes out very close to one, showing that our assertions of linearity are justified (this however only works for apertures that are much larger than how big the stars look at the image).

The resulting fit has RMS of residua 0.07 magnitudes (Fig. 6) and there are no apparent trends when the scans are cut at $A < 8$ (we can now do it for even higher airmasses, thanks to a correction that we discuss later). The zeropoint obviously changes with time and there was a period at the beginning when it was rather fluctuating – we believe that a large effect on the zeropoint changes is in the focusing (again, due to the aperture photometry) and we are planning to implement a temperature compensation for the focusing, which we have already tested on the narrow-field camera. The time series of zeropoints clearly shows the time when the optics was cleaned, as the environment is very dusty. However the effects of the cleaning were reversed rather quickly (Fig. 7).

6 Rayleigh subtraction

To isolate the aerosol information, we must subtract the effects of the molecular atmosphere from the measured extinction. As we have noted, molecular absorption is negligible in the B filter (but very important in the V and R filters) and thus all we need to calculate is the Rayleigh scattering. To do this, we need to characterize both the spectrum of the stars and the spectral response of the FRAM system to form an overall spectral response $R(\lambda)$ (which in general depends on the spectrum of a given star). We considered taking every star to be a black-body radiator of a given temperature, but calculations with realistic stellar spectra have proven this to be a poor approximation (Fig. 8); however, while realistic stellar spectra depend on a number of parameters, the calculated Rayleigh optical depth still correlates well with $B - V$, at least for stars that are not too red ($B - V < 1$), as shown in the left panel of Fig. 9. The spectral properties of the system have been measured in field, but to apparently insufficient precision as for example the rather slight change in the extrapolation of the spectral transmission of the lens shown in red in the right panel of Fig. 9 (with results in pink on the left panel) changes the Rayleigh OD prediction at $B - V = 0$ by 0.01. More precise laboratory measurements are underway.

The Rayleigh scattering as a function of wavelength was then calculated according to [5]. For each scan we take the total atmospheric depth from the GDAS model used at the Auger Observatory [6]. Here it is important to note that simply taking the convolution of the wavelength-dependent Rayleigh VOD with the overall spectral re-

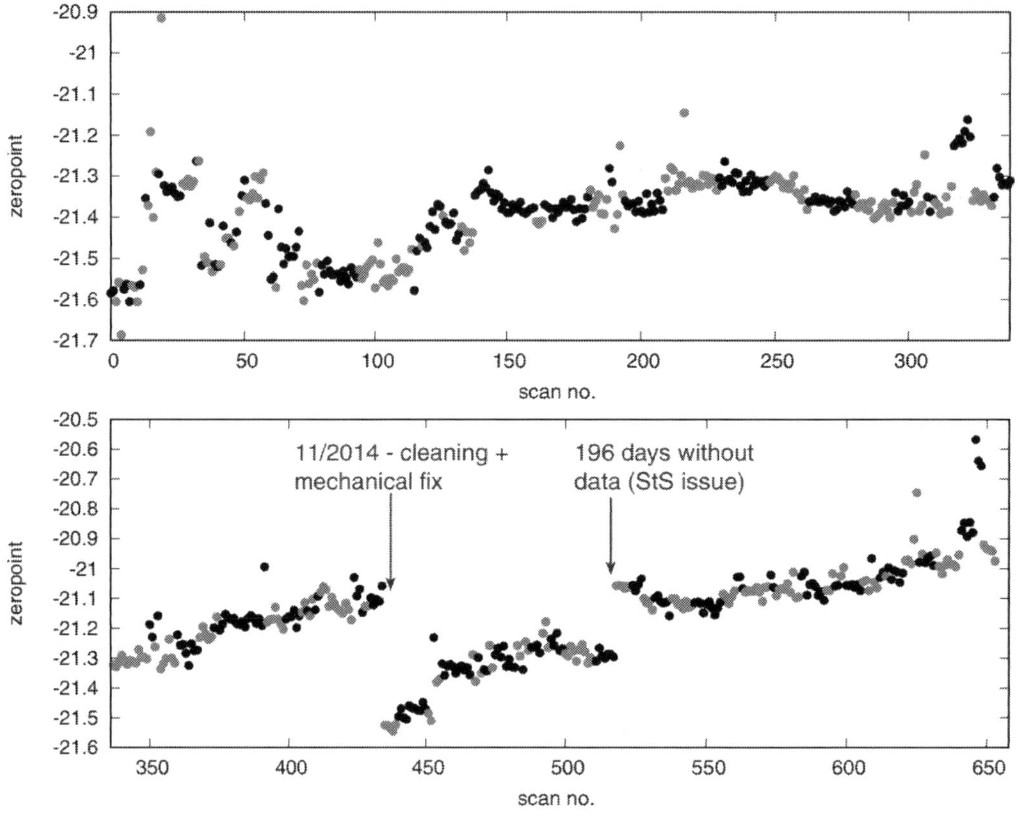

Figure 7. Zeropoint as a function of scan number. Consecutive points with the same color belong to the same night, but there might be large gaps between nights, such as the 196-day gap marked in the bottom row caused by problems with the Shoot-the-Shower trigger system.

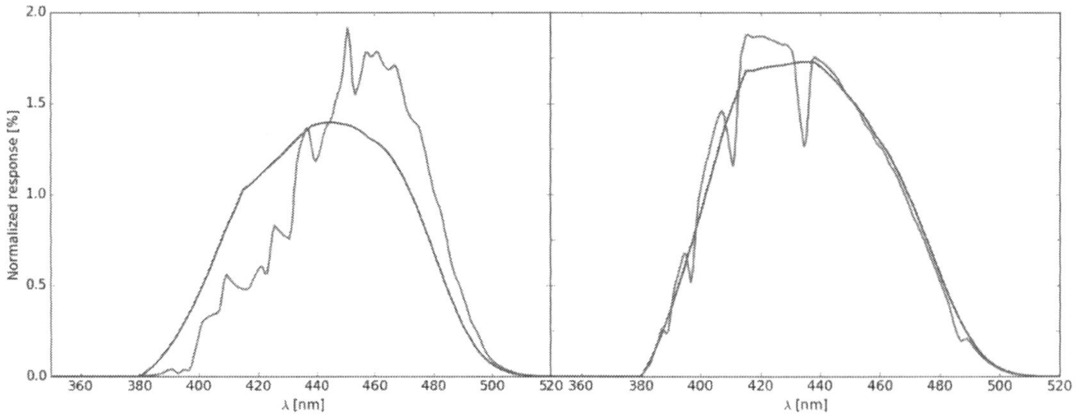

Figure 8. Comparison of the overall spectral response $R(\lambda)$ calculated using black-body radiation (blue) and realistic stellar spectra (green) for stars with surface temperature of 4000 K (left) and 10 000 K (right).

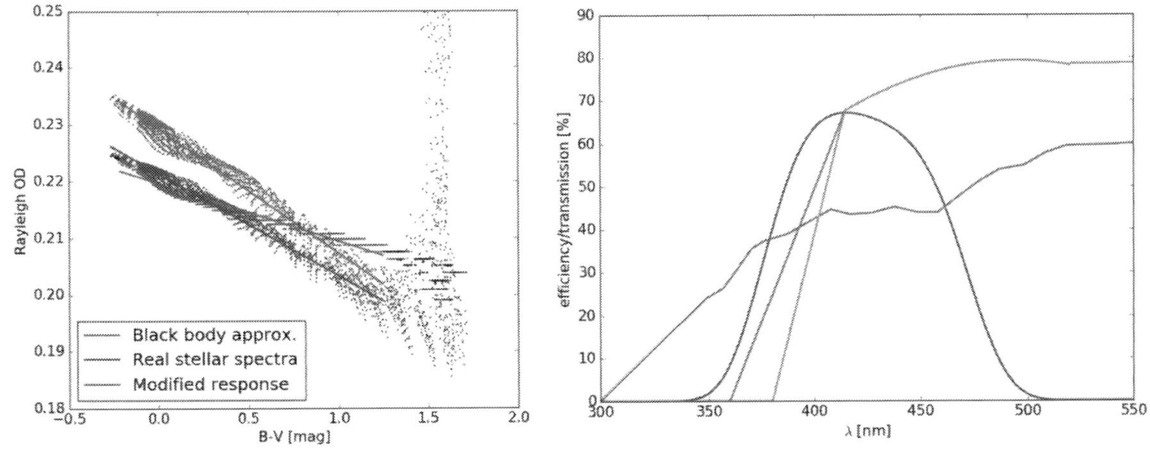

Figure 9. Left: Vertical Rayleigh optical depth at the FRAM site (for typical atmospheric conditions) as a function of (Johnson) $B-V$. The points represent individual spectral models, the lines are linear fits. Shown in green is an approximation based on black-body radiation; a calculation where realistic stellar spectra are used is in blue. The pink points and line show the effect of a modest change in the extrapolation of the spectral properties of the lens into a range of wavelengths where it was not reliably measured; this is shown in the right panel in red, while the original extrapolation is cyan. The green curve in the right panel is the spectral response of the CCD chip and the blue curve shows the transmission of the B filter used.

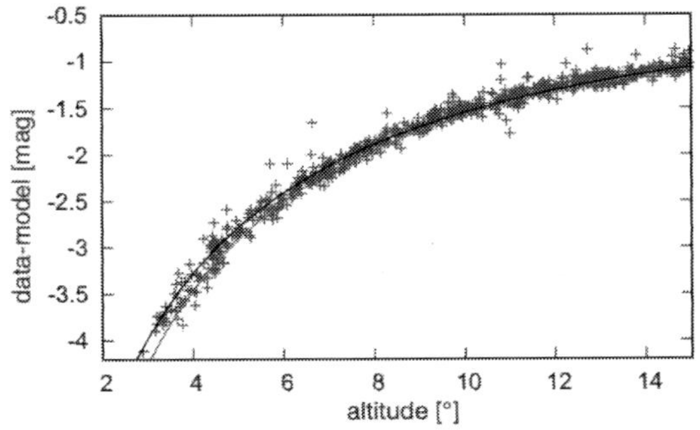

Figure 10. Improvement of the fit (black) taking into account the non-exponential correction due to finite bandwidth in comparison to the naive exponential fit (red).

sponse $R(\lambda)$ is technically wrong, because exponentiation and integration obviously do not commute. Instead, one has to calculate the reduction in the flux at ground for the light across the system passband as

$$\frac{J_{\text{ground}}}{J_{\text{space}}} = \int R(\lambda)\exp\left(-\tau(\lambda)A\right)\mathrm{d}\lambda$$
$$\approx \exp(-\tau_{\text{FRAM}}A + 0.0012A^2). \qquad (15)$$

The $-0.0012A^2$ term is determined numerically and needs to be added to (6) if we want to properly describe stars at $A > 8$ (Fig. 10) and in typical atmospheric conditions $\tau_{\text{FRAM}} \approx 0.208$ for a star with $B-V=0$ with the assumptions on spectral properties of the system and black-body stellar spectra as described above.

The description of stellar spectra can be tested by comparing the color-dependence of the extinction k_C fitted from data (0.015) with the predicted value for Rayleigh scattering, because the extinction is mostly Rayleigh-dominated and the effect of aerosols on k_C is small. Using black-body approximation, the predicted value is too small (0.008), using realistic stellar spectra, the value is slightly too large (0.022); further improvement in this area is clearly needed.[1]

7 Discussion

The presented method allows us to extract VAOD values for scans taken without interference of clouds. In general, we obtain very small but positive values starting from almost zero indicating that the method is working correctly. We also observe that the cleaner nights have the tendency to be stable while dirtier nights show large temporal variations. Most importantly, there is no visible correlation between Z and VAOD, indicating that the fitting procedure does indeed remove this degeneracy and that the variations in VAOD are not artifacts related to the variations of the zeropoint.

The overall precision of the measurement is complex to estimate at the moment. The statistical errors in determination of k from the fits are 0.01 and less. The biggest worry here is whether there could not be a systematic effect with the same trend as the extinction that would influence the fitted value, but the consistency of the fits and the small RMS indicate otherwise. The largest uncertainty then comes from the Rayleigh subtraction as explained above, leading to a possible shift of the VAOD values at

[1] A reader who wishes to study Fig. 9 in detail should note that these results have been calculated using standard Johnson B and V filters, not the filters used in the Tycho2 catalog and thus there is a small difference in the meaning of the $B-V$ values.

the order of 0.01–0.02 depending on how the laboratory measurements of the optics turn out. Moreover, the current calculation of the Rayleigh scattering now overestimates the k_C coefficient: 0.022 is predicted using the spectral properties of the spare lens, while the fitted value is roughly 0.015. Using the theoretical value instead of the fitted one shifts the VAOD by 0.003.

8 Conclusions and outlook

We have shown that we can use the data on apparent brightness of a large number of stars distributed across a wide range of zenith angles gathered in the Shoot-the-Shower program to extract the integral VAOD values to the top of the atmosphere. The methods of data processing, selection of suitable stars and corrections for various instrumental effects lead to a model that fits the observed difference between catalog and measured values with an RMS of 0.07 magnitudes. The model parameters can be then turned into VAOD values using theoretically calculated values for Rayleigh scattering. There are still important questions regarding the use of realistic stellar spectra and the need for detailed knowledge of spectral response of the FRAM components – however these will be addressed with precise laboratory measurements.

The methods to measure the aerosol content of the atmosphere developed with the Auger FRAM have already found application in atmospheric monitoring for the planned Cherenkov Telescope Array (CTA) observatory. See [7] for more details.

Eventually, we would like to establish FRAM as an alternative source of information about the aerosol content of the atmosphere – even though FRAM can provide only the integral value to the top of the atmosphere, it has the advantage of being able to take many measurements independently of the measurements by the fluoresence telescopes and with no interference to their operations. To this end, we have started (since January 2016) to take altitude scans similar to the Shoot-the-Shower data, independently of any StS triggers, during any night when stars are visible above the FRAM site. We are also regularly taking measurements in various azimuth directions in order to study the spatial uniformity of the aerosols and measurements in different filters (B, V and R) in order to measure the Angstrom coefficient every night. The wavelength dependence of the VAOD is usually assumed to be a power law

$$\frac{\tau_\lambda}{\tau_{\lambda'}} = \left(\frac{\lambda}{\lambda'}\right)^{-\alpha} \tag{16}$$

and the mean value of the Angstrom coefficient α has been measured to be 0.7 at the Auger Observatory [8]. However, already these data indicate that the immediate values vary between 0 and 2. Recent data from "nearby" AERONET stations [9] indicate that changes between 0 and 2 can occur during a single day and thus measurement of its value with high temporal resolution is interesting.

Acknowledgments

The authors are grateful for the support by the grants of the Ministry of Education, Youth and Sports of the Czech Republic, projects LG15014 and LM2015038, and Czech Science Foundation project No. 14-17501S.

References

[1] The Pierre Auger Collaboration, Nucl. Instrum. Meth. A **798**, 172 (2015)

[2] Ebr, J. et al., Proceedings of the 3rd Astrorob Workshop, Rev. Mex. AA (SC) **45**, 114 (2014)

[3] The Pierre Auger Collaboration, Journal of Instrumentation **7**, P09001 (2012)

[4] Ebr, J. et al. (The Pierre Auger Collaboration), Proceedings of the 3rd Astrorob Workshop, Rev. Mex. AA (SC) **45**, 53 (2014)

[5] Bucholtz, A., Applied Optics **34**, 2765 (1995)

[6] The Pierre Auger Collaboration, Astropart. Phys. **35**, 591 (2012)

[7] Janecek, P. et al. (CTA Consortium), these proceedings

[8] BenZvi, S. et al. (The Pierre Auger Collaboration), Proceedings of the 30th ICRC, Merida, Mexico

[9] http://aeronet.gsfc.nasa.gov

EPJ Web of Conferences **144**, 01012 (2017) DOI: 10.1051/epjconf/201714401012
AtmoHEAD 2016

FRAM for the Cherenkov Telescope Array: an update

Petr Janeček[1,★], *Jan* Ebr[1], *Jiří* Blažek[1], *Michael* Prouza[1] for the CTA Consortium[★★], *Martin* Mašek[1], and *Jiří* Eliášek[1,2]

[1] *Institute of Physics of the Czech Academy of Sciences, Prague*
[2] *Faculty of Mathematics and Physics, Charles University, Prague*

Abstract. The Cherenkov Telescope Array (CTA) is a project to build a new generation ground-based gamma-ray observatory. The FRAM – F(/Ph)otometric Robotic Atmospheric Monitor – is a device, which can provide it with the Vertical Aerosol Optical Depth (VAOD) maps in the field of view of the CTA in almost real time (delay in minutes). The FRAM is a small fast robotic astronomical telescope with a large field of view and a sensitive CCD camera. It uses stellar photometry to measure atmospheric extinction across the field of view and uses it to calculate the VAOD. In this contribution, results of first few months of operations of the prototype and outlook for deployment at the CTA site will be presented.

1 Introduction

The Cherenkov Telescope Array (CTA) will be the next generation of ground-based very high energy gamma-ray instruments [1]. To meet the strict requirements established for the precision of measurements and efficiency of use of the observation time, it is necessary to have precise knowledge of momentary atmospheric conditions, in particular the presence of clouds and the aerosol content of the atmosphere. Multiple devices are proposed to be installed at the CTA sites [2], one of which is a small robotic optical telescope, the "FRAM" (F(/Ph)otometric Robotic Atmospheric Monitor), which will measure the total Vertical Aerosol Optical Depth (integral to the edge of the atmosphere) in the direction of the CTA field of view using stellar photometry, namely the comparison of measured apparent brightness of stars with respective catalog values [3].

The concept of the CTA FRAM builds on ten years of experience from operation of a similar device at the Pierre Auger Observatory in the Mendoza province of Argentina [4] and the data analysis is directly inherited from the Auger FRAM, with modifications and extensions demanded by the different properties of the CTA Observatory. In particular, the proposed mode of operation is based upon the strategy of using fast scans across a wide range of altitudes above the horizon in order to provide on-demand absolute calibration of the sensitivity of the system (as expressed via the so-called zeropoint Z) and precision global VAOD values simultaneously [5]. Tracking the changes of the zeropoint by virtue of those scans, VAOD maps can be produced from single images of the current CTA field of view with high temporal resolution (1 map per minute) and with only intermittent interruptions (typically 1 calibration scan taking 5 minutes every 30–60 minutes)[1].

A prototype of the CTA FRAM setup has been installed on the grounds of the Institute of Physics in Prague in February 2016, has undergone extensive testing, during which the capabilities to operate in the proposed mode have been demonstrated, and is ready for deployment at a CTA site for preliminary characterization of atmospheric conditions at the site. In this paper we discuss the performance of this prototype and present first steps towards data products for immediate use during the CTA operation.

2 Hardware

The CTA has identified two candidate sites (one at each hemisphere) for the location of the future Observatory: the southern site near Cerro Armazones in Chile and the northern site at the Observatorio del Roque de los Muchachos on La Palma, one of the Canary Islands. Two FRAMs are proposed for the larger southern site (due to spatial variations in weather and parallax with respect to the more distant telescopes), one for the smaller northern site, all three in an identical hardware configuration which consists of a Paramount MYT mount (or similar commercial mount with the capabality to determine its absolute orientation

★e-mail: janecekp@fzu.cz
★★Full consortium author list at http://cta-observatory.org

[1] The actual fequency of calibration scans that will be necessary is currently uncertain and will be determined during the preliminary operation of the FRAMs on their respective locations as it probably depends on the enviromental conditions on the sites as well as on possible future improvements in the method. It is also worth noting that a calibration scan can fail due to presence of clouds that render the fit unreliable; in such a situation, the zeropoint will be extrapolated from previous values and the reported uncertainty on the VAOD in the CTA field of view increased accordingly. Using data from other instruments on site, in particular the all-sky cameras, may allow selection of cloud-free parts of the sky for more reliable calibration.

DOI: 10.1051/epjconf/201714401012

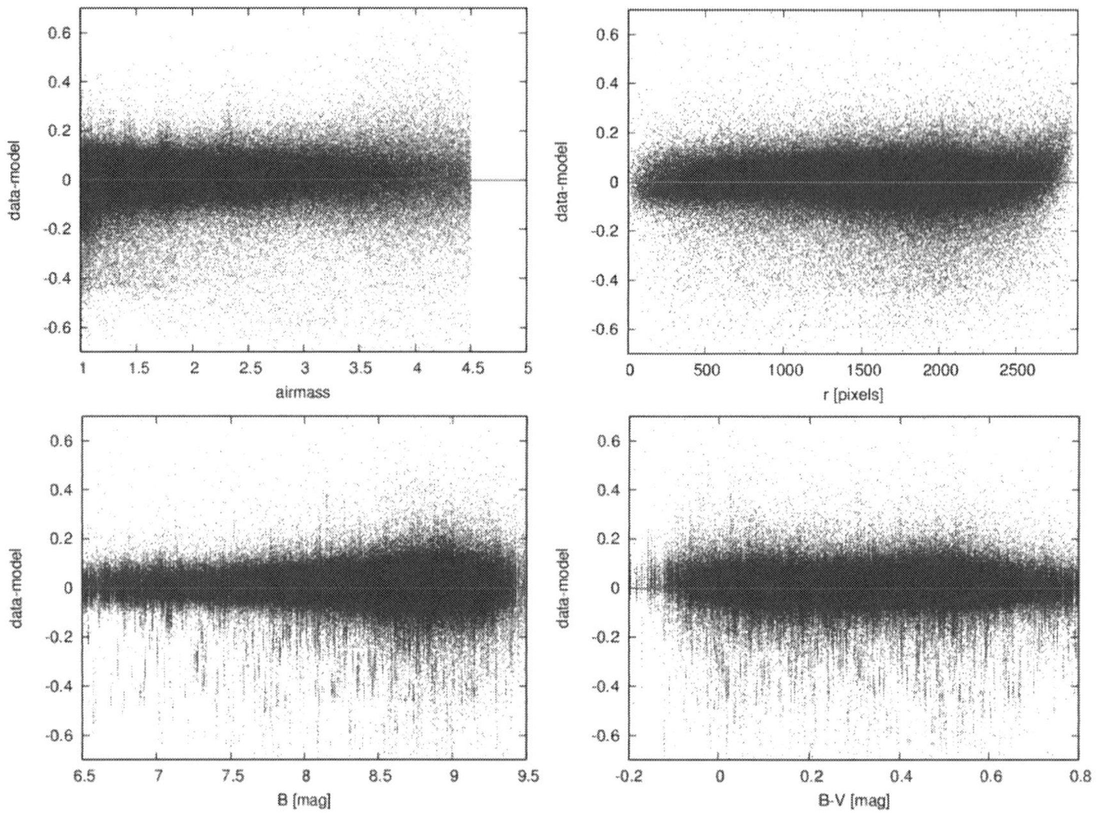

Figure 1. Residua of the global fit as a function of various variables.

after a power cut autonomously), a Zeiss 135/2.0 photographic lens, a Moravian Instruments G4-16000 CCD camera with BVRI photometric filters, a custom enclosure by Elya Solutions and some auxiliary devices. The prototype built in Prague follows the same configuration. The chosen combination of camera and lens provides field of view of 15 × 15 degrees, comfortably covering the whole CTA field of view even when accounting for parallax, while providing almost pin-point sharp star images across the field of view. A single 30-second exposure in the B photometric filter provides hundreds of usable stars anywhere on the sky under reasonable weather conditions.

The whole system is designed for autonomous and safe operation, following our experience with the Auger FRAM, which is located on a remote unmanned location. The custom enclosure was designed in close collaboration with the Elya company. It opens and closes using a hydraulic mechanism and includes fail-safe provisions for the case of a power cut, allowing the dome to be closed using gas pressure stored in a pressure accumulator instead of electricity. Care has been taken to minimize both peak and average power consumption so that the whole system can be operated using a modest solar power plant with 1 kW of installed solar power, 144 kWh of battery capacity and a 2 kW 230 V AC power inverter (needed for the operation of the hydraulic pump) for several days in case of bad weather during the day. This may prove necessary in case the system is deployed on the southern CTA site well ahead of any other infrastructure (the northern site will most likely allow for a power grid connection in any case), as planned in order to characterize the site in advance.

The prototype is installed on a temporary wooden platform fixed into the ground using metal ground screws. This technology has proven useful and it is foreseen that it will be also used during the preliminary-characterization phase, before the FRAM is moved to its permanent concrete foundations. During a site visit at the ORM (La Palma), we have confirmed that the screws can be successfully used in the complicated terrain on the site.

3 Prototype data: altitude scans

The analysis of the calibration scans in altitude proceeds exactly as described in detail in [5]. Here we only remind the reader that the difference between the observed and catalog brightness of a star (in magnitudes) detected during the i-th scan is modeled with the following equation

$$m_{\mathrm{inst}} = m_{\mathrm{cat}} + Z_i + k_i A + c_1(B-V)(c_2(B-V)+1)+ R_1 r(R_2 r+1) + k_C A(B-V) \quad (1)$$

 DOI: 10.1051/epjconf/201714401012

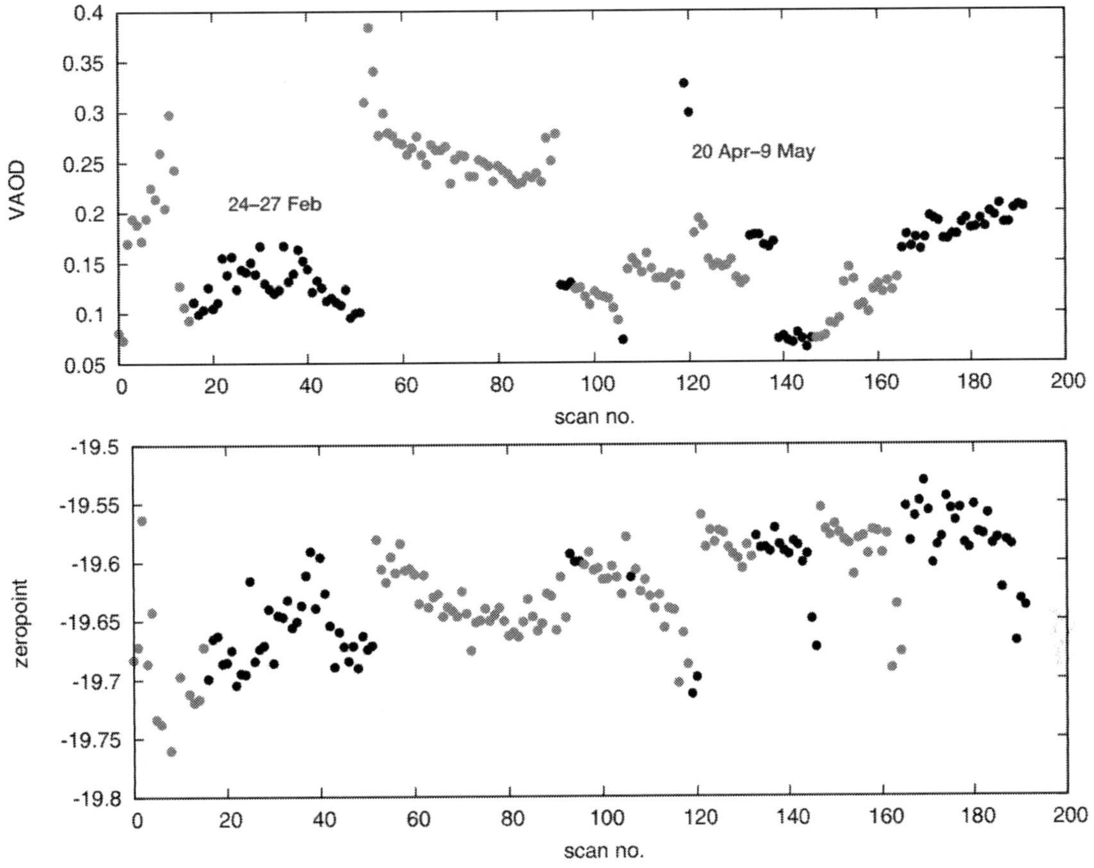

Figure 2. VAOD (upper row) and zeropoint (lower row) as a function of scan number. Consecutive points with the same color belong to the same night, but the nights are not neccessarilly consecutive. The first three nights are between 24 and 27 February, the rest between 20 April and 9 May – the very high VAOD values on 27 February are not surprising, as winter nights are mostly foggy in Prague.

where Z_i is the zeropoint (calibration constant) for the scan, k_i is the extinction coefficient (1.086× the vertical optical depth), A is the airmass in units of the vertical airmass, B and V are magnitudes of the star in the catalog in the respective filters, r is the distance of the star from the center of the respective image and M, c_1, c_2, R_1, R_2 are parameters of the system (optics+CCD+data reduction) and the last term represents the dependence of the extinction on the star color.

This analysis can proceed only when the scan is not affected by clouds, which is a rather rare condition in Prague, in particular in winter – we collected data in 12 nights between February and May 2016, totaling almost 200 usable scans. Fig. 1 shows the residua of the global fit over all those scans. Only stars with $A < 4.5$ could be used because of trees and other disturbances near the horizon. The overall RMS is 0.09, slightly larger than in the Auger FRAM data analysis (0.07). This expected as the conditions in Prague are much worse, mainly when it comes to background and parasitic light and the analysis is not yet optimized for the different hardware setup of the CTA FRAM prototype.

Fig 2 shows the results for the zeropoint and VAOD in time series. Unlike in [5], here only an approximate Rayleigh model with black-body spectra has been used; however as the aerosol content in the urban atmosphere is rather high, this is not a significant effect. Note that the VAOD changes heavily even within a single night, but mostly in a smooth fashion. The zeropoints both drift and fluctuate. In order to obtain zeropoint for a given time within a night for which successful calibration scans are available, we fit a smooth function (quadratic in time) to the zeropoint values for each individual night. The spread of values around the fit is 0.02 mag and constitutes the main contribution to the uncertainty of the determination of VAOD maps described below. We suspect that the fluctuations of the zeropoint are related to thermal contraction of the lens with decreasing temperature during night, but this effects needs to be investigated more. The long-term drift of the zeropoint might be due to dust accumulation.

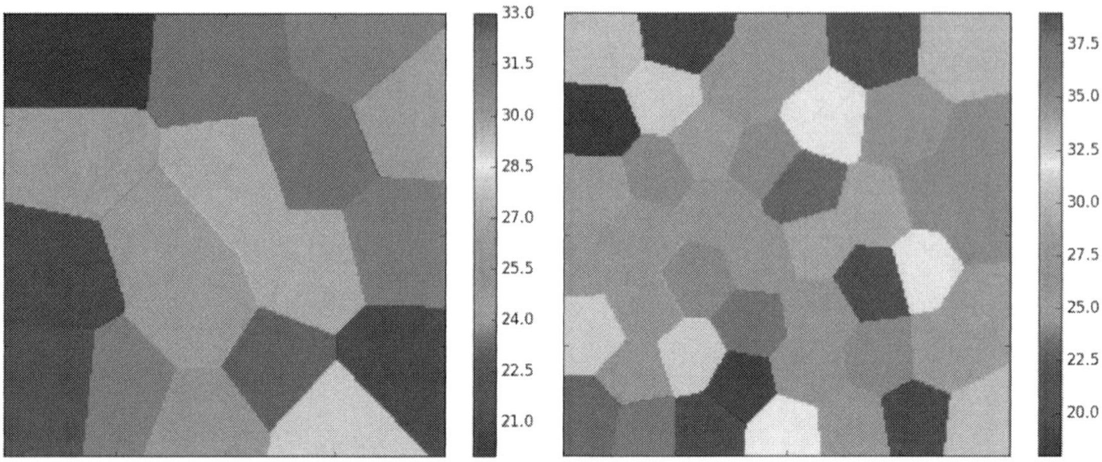

Figure 3. Examples of Voronoi tessellations for a poor (left) and rich (right) star fields. The color scale indicates the number of stars in each cell.

4 VAOD maps

For a night for which a zeropoint calibration is available, we can determine the VAOD at any point of the sky from a single image. We have taken series of images of fixed fields of view in between the altitude scans to simulate the proposed mode of operation of the FRAM within the CTA.

To produce a VAOD map from an image, we must first perform 2D binning of the stars into regions, because a measurement of a single star is too imprecise. As the density of stars varies greatly across the sky, this calls for an adaptive algorithm. We have chosen the so-called Voronoi relaxation which ensures that each spatial bin will contain between k and $2k$ stars for a chosen value of k [6]. A Voronoi tessellation is a set of regions defined by a set of central points as areas that are closer to a given point than to any other of the points. The Voronoi relaxation algorithm starts with a regular grid. For each cell the center of mass of stars contained within the grid is calculated and centers of mass of all grids form a set of points that defines the first iteration of the Voronoi tesselation. Now the algorithm proceeds iteratively until it converges. In each step, cells with less than k stars are dropped and cells with more than $2k$ stars are split into two and new centers of mass are calculated. (In a real-world implementation, the procedure is slightly more complex for optimization reasons). Note that the tesselation is chosen based on the catalog, that is the stars we expected to detect in the given field of view. In the presence of small clouds (or even thick aerosols), some stars might not be detected in some parts of the field of view. In such a case, the adaptive algorithm applied to the detected sources would create large regions around these areas to compensate for the lack of stars, effectively concealing the presence of the clouds.

When the 2D bins are defined (using $k = 20$), we calculate the local VAOD as the average of

$$0.921\frac{m_{\mathrm{inst}} - m_{\mathrm{model}}}{A} - \tau_{\mathrm{Rayleigh}} \qquad (2)$$

over the stars in the bin. Assuming that the RMS of the global fit reflects mainly the individual uncertainties in measured and modeled brightness of the stars, the statistical uncertainty of an average over at least 20 stars is smaller than 0.02. Combined with the uncertainty in the zeropoint calibration of 0.02, the overall uncertainty of each cell in the VAOD map is roughly 0.03. This is still rather large for an absolute VAOD measurement, but when viewing sequences of such maps over time, the appearance of even thin cirrus cloud is immediately obvious. Some examples of VAOD maps are shown in Fig. 4

It should be noted that Prague serves as a worst case scenario for this measurement, because the sky is both bright and unevenly illuminated, parasitic light is omnipresent, accessibility of the horizon and thus the lever arm in the model fit is limited and moreover, the extinction is dominated by aerosols (not the Rayleigh scattering) and thus there are additional problems with the determination of the dependence of the extinction for a single star on its color index. All of these problems will disappear once the system is deployed on a dark and clear site, such as those selected to host the CTA.

5 Summary and outlook

The FRAM prototype has demonstrated its ability to measure VAOD in a given field within self-calibrated operation and is ready for deployment. There are however many tasks for future work. On the side of analysis, we have yet to optimize catalog matching and photometry, improve model description of the data and understand the changes of the zeropoints with time. On the hardware side, we still need to deploy the prototype device on site and build two more FRAMs. Finally, the most difficult task will be to integrate the software with CTA array control and data processing and, of course, operate the FRAMs throughout CTA's lifetime.

DOI: 10.1051/epjconf/201714401012

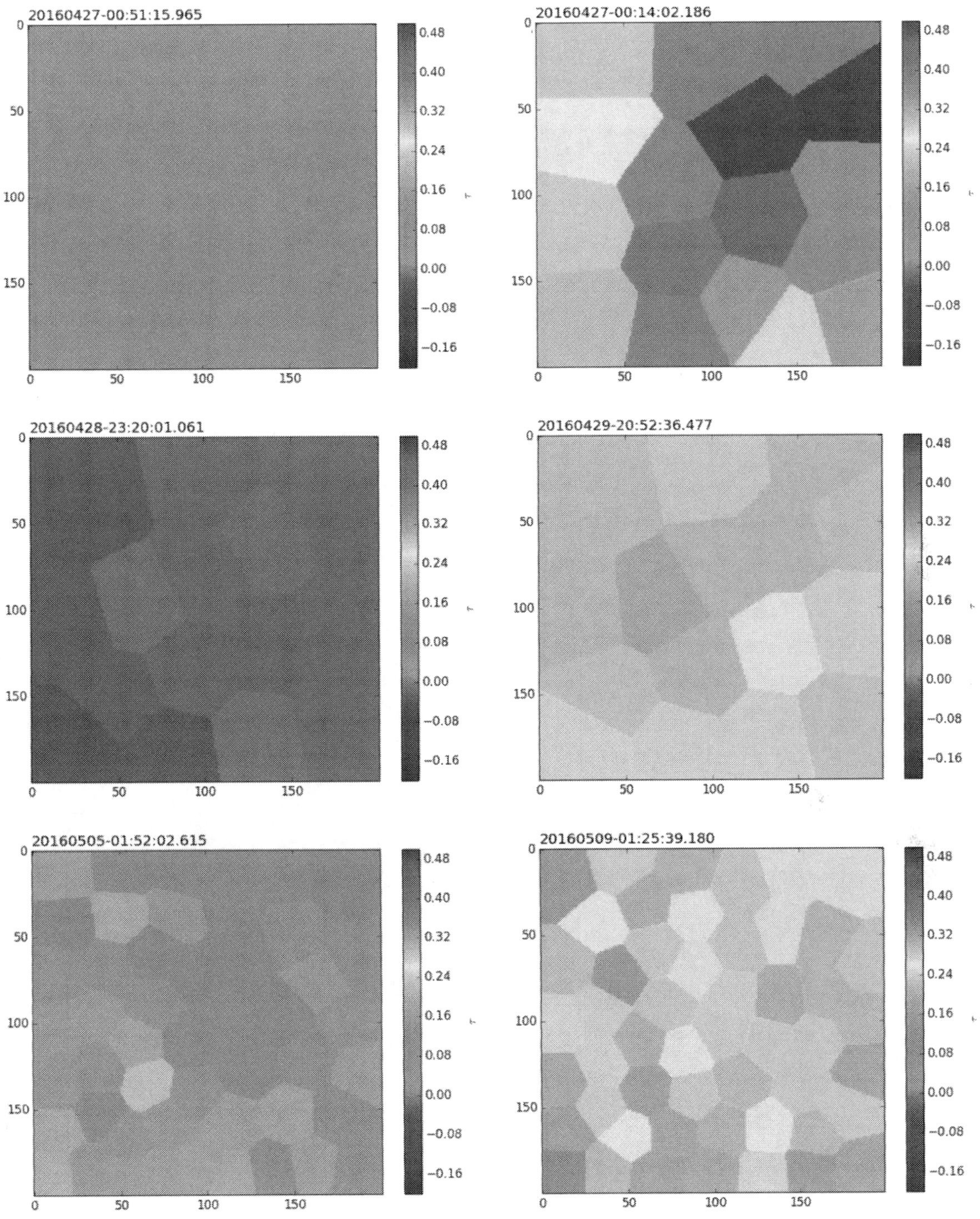

Figure 4. Example VAOD maps under varying weather conditions. X and y coordinates are in arbitrary units, the image plotted area corresponds to the whole 15 × 15 degree image. The upper left image shows very clear conditions while the upper right image shows a thin cloud, which allows the detection of most expected stars, but causes clear inhomogeneity in VAOD across the field of view. The rest of the images illustrate the various kinds of conditions encountered in Prague during the prototype operation.

Acknowledgments

We gratefully acknowledge support from the agencies and organizations listed under Funding Agencies at this website: http://www.cta-observatory.org/. The authors are grateful for the support by the grants of the Ministry of Education, Youth and Sports of the Czech Republic, projects LE13012, LG14019 and LM2015046.

References

[1] Acharya, B.S. et al. (CTA Consortium), Astroparticle Physics **43**, 3–18 (2013)

[2] Gaug, M. et al. (CTA Consortium), these proceedings

[3] Ebr, J. et al. (CTA Consortium), Proceedings of the 34th ICRC, The Hague, The Netherlands

[4] The Pierre Auger Collaboration, Nucl. Instrum. Methods Phys. Res., Sect. A **798**, 172 (2015)

[5] Ebr, J. et al. (The Pierre Auger Collaboration), these proceedings (2016)

[6] Du, W., Faber, V., Gunzburger, M., SIAM Review **41(4)**, 637 (1999)

EPJ Web of Conferences **144**, 01013 (2017) DOI: 10.1051/epjconf/201714401013
AtmoHEAD 2016

New atmospheric profiles for H.E.S.S. data analysis

Mónica Seglar-Arroyo[1,*] and *Fabian* Schüssler[1,**] on behalf of the H.E.S.S. Collaboration

[1] *IRFU, CEA Saclay*

Abstract.
Particles arriving to Earth allow us to scan and improve our understanding of the Universe, both in its composition and its dynamics. For ground-based experiments in astroparticle physics, the atmosphere plays a key role and its understanding, monitoring and modelling is essential for a realistic description in data analysis and simulations. In this work we present the development of a novel experimental atmospheric model, the GNNA120 model, describing the atmosphere above the H.E.S.S. gamma-ray observatory in Namibia. This new description is the most realistic to date and introduces the possibility to study variations at different time scales. This enables us to provide the first ever study of seasonal effects based on actual measured atmospheric profiles for the H.E.S.S. observatory and opens up the window to further improvements of the H.E.S.S. Monte Carlo simulations.

1 Introduction

Much of the radiation propagating in the Cosmos and incident on the Earth is thermal radiation generated in hot objects such as stars. However, it is well-known that certain particle populations in the Universe cannot result from thermal processes, but they must be produced by collective mechanisms. The best-known of this type of population are cosmic rays.

Even though the existance of cosmic rays has been known since the first decades of the 20th century, their acceleration locations have not been found yet. In fact, deflections of charged particles in galactic and/or extragalactic magnetic field are so large that over most of the energy regime, the pointing information is completely lost; only at the very highest energies, where the energy lowers the deflection effect, in the domain of the AUGER experiment, the directional information can possibly be exploited. In this context, non charged particles neutrinos and photons play a key role to study high-energy cosmic rays and their sources.

The detection of gamma-rays can be done on board of a satellite, e.g. Fermi-LAT or AMS, or at ground level, e.g. Veritas, MAGIC and H.E.S.S., depending on the energy range. The atmosphere acts as an active target in ground-based experiments, since it is the target medium for the cosmic particle, but also the emitter of Cherenkov photons and the transport medium for those photons. Being a continuously changing system, constantly monitoring and correct parametrisation of the atmosphere plays a crucial role. Varying atmospheric conditions in terms of state variables may alter the development and detection of extensive air showers. The main goal of the study presented here is to revise the H.E.S.S. atmospheric parametrisation and quantify the systematic errors from an incorrect description.

The paper is structured as follows: In §2,the characterisation of the atmosphere above the H.E.S.S. site is presented. Following this, the new experimental GNNA120 atmospheric model is introduced and explained. In §3, we present and compare the former and new atmospheric profile and other atmospheric features. The results are discussed and summarized in §4.

2 Characterising the atmosphere above H.E.S.S.

2.1 Previous studies of the atmosphere above H.E.S.S.

H.E.S.S. is a stereoscopic system consisting of five imaging atmospheric Cherenkov telescopes located at 1800 meters above sea level in the Khomas highlands of Namibia (23°16' 18" South, 16° 30' 00" East). The four H.E.S.S. phase I telescopes are arranged in a square of 120 side length, with the large H.E.S.S. II telescope at its centre.

In the H.E.S.S. experiment, the gamma-rays are detected indirectly via the Cherenkov light emitted by the charged particles of the induced air-shower. Thus, a good understanding of the atmosphere that acts as the calorimeter of the system is crucial for the Imaging Atmospheric Cherenkov Technique (IACT) method.

In 1999, motivated by the advent of a new generation of gamma-ray observatories, atmospheric density profiles, as well as several light absorption and scattering processes dependency on geographic position and time variability were investigated by Konrad Bernlöhr [1]. The most important aspects treated in the mentioned work are:

*e-mail: monica.seglar-arroyo@cea.fr
**e-mail: fabian.schussler@cea.fr

- **Vertical structure of the atmosphere.** Being the most influential parameters, it was seen that different density profiles lead to differences in Cherenkov light density up to 60%. Concretely, seasonal variation at mid-latitude sites were found to be of the order of 15-20% as can be seen in Figure 1 where several average lateral distributions of Cherenkov light photons have been simulated with `CORSIKA 5.71` [2] taking into account absorption of Cherenkov light.

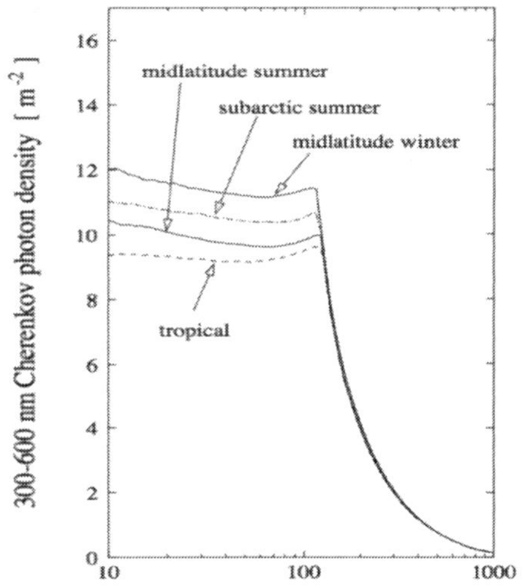

Figure 1: Average lateral distribution of Cherenkov light photons in the wavelenght range 300-600 nm for vertical 100 GeV gamma-ray showers in `CORSIKA 5.71` simulations with different atmospheric profiles (200 showers) from [1]

- **Atmospheric extinction of Cherenkov light.** It includes absorption and scattering of the electromagnetic radiation by dust and gas between the emitter and the observer. For those phenomena, the main concern is that extrapolation for high-altitudes must be avoided, and monitoring procedures must be applied even though a measurement of the aerosol vertical structure is rather difficult. However, Lidar-based methods are available to measure the extintion profile with a 10% accuracy.

  Note that within the H.E.S.S. data analysis and simulation tools, the *Cherenkov transparency coefficient* [3] takes into account the presence of elevated aerosols concentrations and of large-scale clouds. The parameters that are used as an input for its calculation are the energy threshold of the telescope, which shows a direct proportion to the transparency of the atmosphere, and other inversely proportional quantities related to the telescope data taking efficiency like the telescopes-wise muon efficiency, average pixel gains and number of active telescopes. This approach has been proven to be an effective and a pratical method of distinguinshing good quality datasets without the need to measure the aerosols distributions [3] , that is a rather difficult task.

- **Refraction.**The wavelenght independency of the refractive index was assumed since n(λ)-1 changes only 5 % over the wavelenght range 300-600 nm, which is the range typically covered by photomultipliers. The influence on gamma-ray observations can therefore be considered as negligible [1].

The vertical atmospheric profiles used within KASKADE [4], one of the two main simulation frameworks used within the H.E.S.S. collaboration, was derived from balloon soundings at Windhoek airport in the 1990s, averaged over a whole year. A verification of this parametrisation is the main aim of the work presented here.

2.2 Experimental GNNA120 atmospheric model

Worldwide, the atmosphere is being continuously monitored by a fleet of satellites, networks of weather stations, ballon soundings, etc. Here we exploited different databases belonging to NOAA and NASA's monitoring and modeling centers, respectively. A full atmospheric profile model, named experimental GDAS - NRLMSISE for Namibian Atmosphere up to 120 km (GNNA120) model is introduced. For this new description of the atmosphere above H.E.S.S. two different sources of data have been chosen.

- **Low altitudes.** The Global Data Assimilation System [5] (GDAS) is a system used by the National Center for Environmental Prediction (NCEP) Global Forecast System (GFS) to place observations into a gridded model space for the purpose of starting weather forecasts with observed data. The full GDAS data are publically available via the NOAA National Operational Model Archive and Distribution System (NOMADS) [6] which is a Web-services based project providing both real-time and retrospective access to climate and weather model data.

  GDAS adds various types of observations to a gridded, 3-D, global model: surface observations, ballon data, wind profiler data, aircraft reports, buoy observations, radar observations, and satellite observations. Based on simulations and weather predictions tied to measurements model data is available with a time resolution of 3 hours. The provided data contain information of 23 different levels, up to an altitude of about 25 kilometers above the surface. The levels are related to the pressure at the layer and span a range from 1000 hPa to 20 hPa. The GDAS grid spans the whole Earth with a grid-size is of 1°x1° in longitude and latitude.The data available in every dataset are temperature, pressure, geopotential height, u-wind, v-wind, vertical pressure velocity and relative humidity.

  The coordinates for the HESS II experiment (23.27 ° S, 16.5 ° E , and 1800 m ASL, Namibia) while the closest station that launches weather ballons is located at the Windhoek airport (22.57 ° S, 17.10 ° E, and 1700 m ASL, Namibia).

- **High altitudes.** The Community Coordinated Modeling Center (CCMC) provides several empirical models.

The NRLMSISE-00 Model 2001 [7] describes the neutral temperature and densities for the Earth's atmosphere up to thermospheric heights. It includes temporal and spatial variatons at various scales based on an analytical model.

The profiles that could be extracted from GDAS only present data up to around 25 km. Thus, once one runs out of data from the GDAS database, the data is taken from the NRLMSISE-00 model up to 120 kilometers, in layers of 5 kilometers. It has been verified that both models agree very well in the overlapping region, thus no artefacts are induced by this combination.

3 Results

3.1 GNNA120 atmospheric profiles

Temperature and density profiles for the combined data set extracted from GDAS and NRL-MSISE are shown in Figure 2. One can appreciate the expected changes of the profiles while increasing the altitude, from troposphere up to thermosphere, going through stratosphere and mesosphere. The steep changes in temperature correspond to the pauses between these layers.

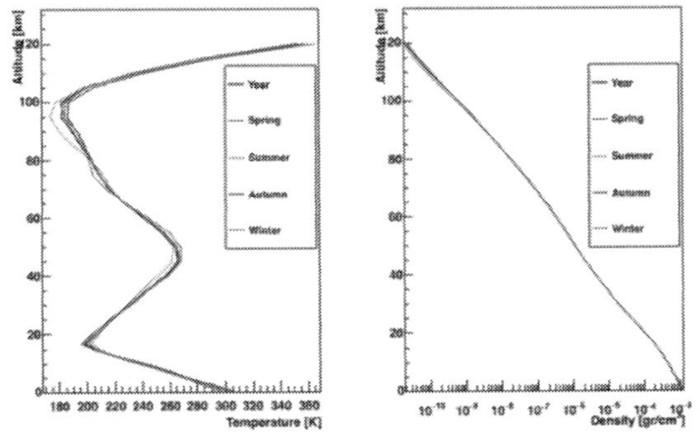

Figure 2: (a) GNNA120 yearly and seasonal averaged temperature profiles up to 120 km (b) Yearly and seasonal averaged density profiles up to 120 km.

In order to check the impact of the humidity on the density profiles, the comparison for the averaged density profiles while taking into account humidity effects is plotted in Figure 3 for the different season and the whole year. For computing the humidity, a very accurate formula for determining the saturation vapor pressure by Herman Wobus in [8] has been used. The relative changes have a maximum value of 0.7%, what lead us to the conclusion that this effect is rather small compared to other systematics.

Note that the humidity is an important parameter for fluorescense techniques and that in the case of condensed water, it is currently taken into account into the transparency coefficient. It can also be noted that in future experiences as CTA, it would be measurable by Raman LIDARs, and included in aerosols simulations.

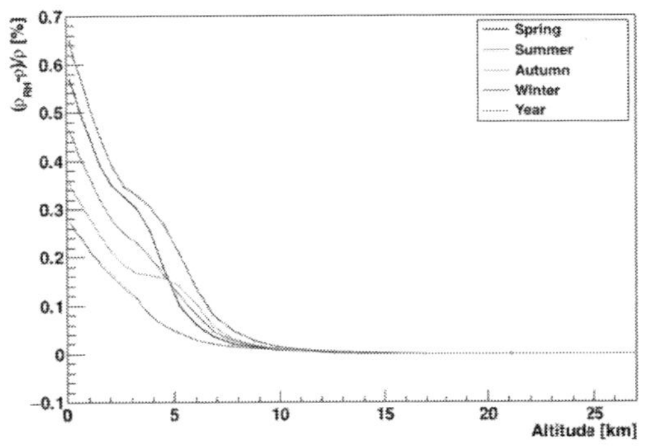

Figure 3: Difference in percentage of the density averaged profile when taking into account humidity effects

3.2 Comparison KASKADE-to-GNNA120 atmospheric profiles

In order to compare the former and the new GNNA120 atmospheric profiles, relative residuals have been computed and shown in Figure 4.

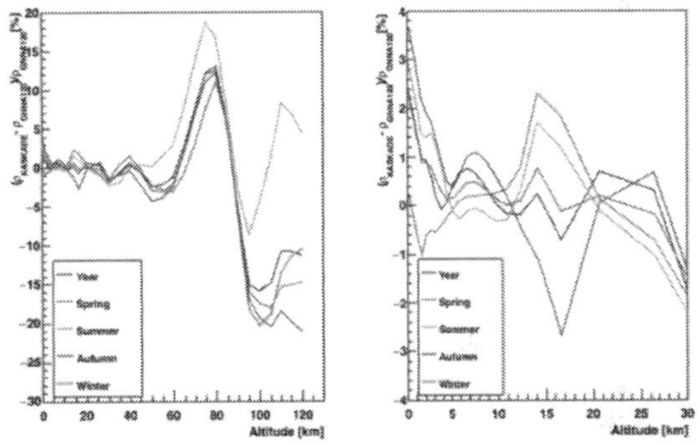

Figure 4: (a) Residuals of the computed GNNA120 model and the H.E.S.S. profiles used for simulations (b) Residuals zoomed at the ROI, important for the gamma-ray induced air shower developement.

The biggest changes (up to 20%) are found in the upper regions of the atmosphere. However, the region of interest (ROI) important for the development of TeV gamma-ray induced air showers ranges from an altitude of 2 km to 26 km. There, the largest differences, up to 4% from the old KASKADE profile, are given for the seasonal GNNA120 profiles, namely for spring and autumn, around altitudes of 15-20 kilometers. Comparing KASKADE to the yearly GNNA120 profile, the changes are less significant and do not exceed 1%.

3.3 Slant depth

In principle, one can assume that the big changes up to 20% seen in Figure 4 for the density in the upper layers of the atmosphere will not change the development of the shower in simulations, since the density goes rapidly to zero as one increases to these altitudes. Seeking to test that assertion, we calculate the cumulative slant depth since it gives an estimation of the changes of the place in the atmosphere where the first interaction takes place and the following air shower evolution. The yearly-seasonal comparison is shown in Figure 5 where the biggest changes up to 3 g/cm^2 are visible for autumn and spring.

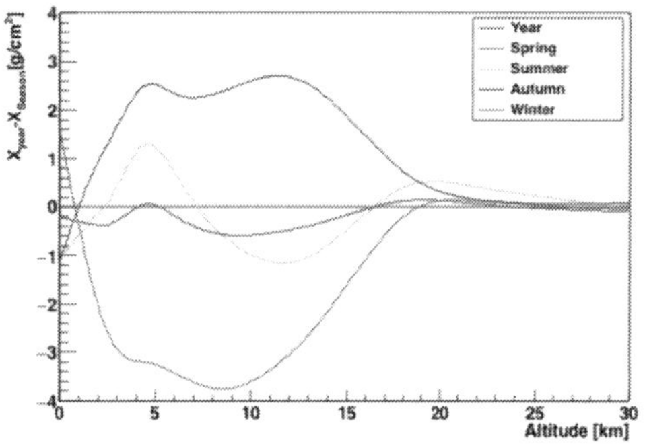

Figure 5: Differences between seasonal averaged and yearly averaged slant depth as a function of the altitude in the ROI. The altitude range was set to the ROI since for high altitudes, it goes rapidly to zero.

We note that the size of the effect is small compared to the intrinsic shower-to-shower fluctuations.

4 Summary

The impact of the parametrisation of different atmospheric variables relevant for the atmospheric Cherenkov technique in the atmosphere above the H.E.S.S. site has been presented by means of data coming from GDAS and NRLMSISE, the new GNNA120 model has been built.

The atmospheric profile used in the H.E.S.S. analysis, namely a yearly average at four different altitudes with interpolated values for the density up to 120 km, has been compared to the new GNNA120 model. In the ROI, the yearly comparison to the KASCADE profile shows differences of about 1%, meaning that the implementation of the yearly averaged atmosphere in the H.E.S.S. analysis is fully acceptable. This implies that, overall, we found surprisingly small differences with the 20 year old implementaton in KASKADE and, thus, our study can therefore be considered as a validation of the profiles used in the H.E.S.S. analyses so far.

Note that the GNNA120 model including seasons, developed during this work, has been included in the HESS analysis software. Results show up to a 20% of difference and the following air shower evolution at high altitudes and up to 3% for the seasonal profiles in the important range for the gamma-ray induced air shower development. This emphasizes the importance of including seasonal changes and encourages us to further study effects at even shorter timescales.

The possibility of including humidity effects into the density profiles has been studied. The effect never exceeds 1% in any of the studied cases. Despite this fact, it is a parameter that presents huge spatial and temporal fluctuations. Thus, we decided to not take it into account in the average profiles.

Some changes with respect to the yearly slant depth profile have been found for autumn and spring. The scale of these (<4 g/cm2) is modest and should not significantly influence the air shower simulations.

After checking the impact of the seasonal changes, one wants to go one step further by taking into account small-scale effects. The presented study opens up the window to further implementations of the parameters that describe the chaotic nature of the calorimeter of our experiment, the atmosphere. The next step of the GNNA120 model and, indeed, the ultimate goal would be to implement atmospheric profiles every 3 hours a new method for Monte Carlo simulations: : simulations reproducing the actual data taking conditions in a time dependent fashion, so called Run-Wise Simulations currently under development within the H.E.S.S. Collaboration.

References

[1] Bernlohr, Astroparticle Physics **12 (4)**, 255–268 (2000)

[2] D. Heck, J. Knapp, J. Capdevielle, G. Schatz, T. Thouw (https://www.ikp.kit.edu/corsika/) Wissenschaftliche Berichte, Forschungszentrum Karlsruhe **FZKA 6019** (1998)

[3] Hahn, Joachim and de los Reyes, R, EPJ Web of Conferences **89**, 02002 (2015)

[4] M.P. Kertzman and G.H. Sembroski, NIM A **343**, 629 (1994)

[5] Global Data Assimilation System, https://www.ncdc.noaa.gov

[6] NOAA National Operational Model Archive and Distribution System, http://nomads.ncdc.noaa.gov/

[7] Community Coordinated Modeling Center, http://ccmc.gsfc.nasa.gov/modelweb/atmos/nrlmsise00.html

[8] Wahiduddin, https://wahiduddin.net/calc/density_altitude.htm

EPJ Web of Conferences **144**, 01014 (2017) DOI: 10.1051/epjconf/201714401014
AtmoHEAD 2016

Overview of Atmospheric Simulation Efforts in CTA

Michal Vrastil[1,2,*]

[1] *Institute of Physics of the Czech Academy of Sciences, Na Slovance 1999/2, 182 21, Prague*
[2] *Charles University, Faculty of Mathematics and Physics, Ke Karlovu 3, 121 16, Prague*

Abstract. The Cherenkov Telescope Array (CTA) is an observatory for ground-based gamma-ray astronomy currently under construction, which will observe photons with very high energies (20 GeV – 300 TeV). One of the main contributions to the systematic uncertainties stems from the uncertainty on the atmospheric density profile, of molecules and aerosols. To minimize these systematics a full calibration of the atmospheric properties is important as well as a calibration of the detector response. In the paper we introduce the strategy for atmospheric simulations, use of Monte Carlo simulations and available CTA computing resources. We also describe in more detail realized and planned atmospheric simulations as well as the Czech contribution to this effort.

1 Introduction

The Cherenkov Telescope Array (CTA) is the next-generation instrument for detecting very high energy (VHE) γ-rays (20 GeV – 300 TeV) and it will be the major global observatory for gamma-ray astronomy [1]. These VHE γ-rays originate from non-thermal processes in the universe, both galactic (supernova remnants, pulsars or pulsar-wind nebulae) and extragalactic (active galactic nuclei, starburst galaxies, galaxy clusters or gamma ray bursts). VHE γ-rays cannot be detected directly on the Earth. Instead, when reaching the Earth's atmosphere they interact with atmospheric nuclei and electromagnetic showers start to develop. Particles in the shower (mostly electrons and positrons) emit Cherenkov light (mainly in near UV and optical band) which reach the ground level and can be detected by Imaging Atmospheric Cherenkov telescopes (IACTs).

The development of the shower and propagation of the Cherenkov light have great impact on the performance of the observatory (in terms of sensitivity, energy and angular resolution, effective area of arrays, etc., for more details about performance see [2]). The performance is influenced by many things such as atmospheric conditions, geomagnetic field, altitude of the site or night-sky background. All of these effects need to be studied by means of detailed Monte Carlo (MC) simulations. For a general overview of the influence of atmospheric conditions on the performance of an IACT, and the corresponding atmospheric calibration strategy for the CTA, see [3].

2 Monte Carlo Simulations

Detailed MC simulations of the detector response to extensive air showers are widely used within the CTA collaboration. These simulations are used e.g. for characterization of performance of the CTA, almost any science analysis in the CTA (via the generation of instrument response functions), optimization and validation of system parameters during prototyping and the (pre–)construction phase, determination of upgrade scenarios, development of reconstruction algorithms or evaluation of systematic uncertainties.

For these tasks, the CTA MC working group studies the impact of various parameters with large sets of simulated events for different primary particles: protons, electrons and γ-rays in the energy range from 3 GeV to 300 TeV. For example, during the site selection, nine sites (three in the northern and six in the southern hemisphere) with different altitudes, geomagnetic fields, night-sky background and atmospheric conditions were studied with MC simulations in order to determine their scientific performance.

For the simulation of the Cherenkov light emission the CTA collaboration uses the CORSIKA program [4]. CORSIKA was chosen as the basis of the air shower simulations due to its public availability, its rich set of interaction model options and the fact that it is used and tested by a large number of experiments. The simulation of the detector response is usually done by a second program called `sim_telarray`[5], which was originally developed for simulating the HEGRA telescope system. During years the program became much more configurable and flexible.

After simulations of showers and telescope responses these data have to be processed. For this analysis several independent analysis tools can be used: Eventdisplay [6], MARS [7] or FAST[1] [8]. Results of these analyses are then used for CTA's needs.

[1]FAST does not use `sim_telarray` simulations as input, but performs a simplified analytical calculation of the detector response.

*e-mail: vrastil@fzu.cz

 DOI: 10.1051/epjconf/201714401014

2.1 CTA Computing Resources

MC simulations (and other simulations) for the CTA are performed on the CTA Computing Grid (CTACG). CTACG uses the European Grid Infrastructure (EGI) grid [9] through the CTA Virtual Organization since 2008. EGI provides access to resources across Europe using grid computing techniques. The CTA has access to twenty EGI sites in 7 countries and one Advanced Resource Connector (ARC) site in Sweden. Every CTA member can get access to these resources through the CTA Virtual Organization. To access all these resources the CTA is using the DIRAC (Distributed Infrastructure with Remote Agent Control) Interware [10]. DIRAC is a software framework for distributed computing which provides a solution for workload and data management system. More information about DIRAC setup for the CTA can be found in [11].

CTACG available resources are on average 6000 – 8000 (peak) cores while 1000 years of CPU time are spent each year. Available storage is about 2.3 PB of disk and 0.7 PB of tape storage which are distributed among six main sites [12]. It turns out that the storage is usually the main problem of simulations – disks are filling with data very rapidly during CTA mass productions (which are required by the CTA Consortium).

3 Atmospheric Simulations

As mentioned earlier the Earth's atmosphere is very important for any IACT – we are using the atmosphere as our calorimeter and therefore it is an integral part of the detector. Atmospheric quality has a great impact on the air-shower development itself, the variation of the Cherenkov angle with altitude, the loss of photons due to scattering and absorption of Cherenkov light out of the camera field-of-view (see [3]).

For these reasons we need to have precise measurements of the atmosphere such as detailed knowledge of the molecular profile, its density and refractive index, ozone absorption and aerosol extinction of Cherenkov light. To get this information CTA plans to use several instruments measuring the atmosphere — Raman LIDARs, the FRAM telescope, All-Sky-Cameras, Ceilometers, sun and moon photometers and radiosondes (see e.g. [13], [14], [15] or [16], for an overview see [3]).

In our study we focused on two main aspects of the atmosphere — molecular density profile and the presence of aerosols — and their effects on the estimated performance of the Northern observatory, i.e. effective area, energy bias and resolution of the array, and angular resolution. We studied several different atmospheric profiles (density, thickness, refractive index, and optical depth profiles), see [17]. Figure 1 shows different extreme density profiles with two reference profiles `Average Winter` and `Average Summer`. Figure 2 shows these profiles in comparison with the average profile `Average Winter`. From Figure 2 we can see what properties different extreme profiles have — e.g. `Extreme14.0_low` (green line) describes an atmospheric profile with density minimum (relative to the `Average Winter`) at an altitude of 14 km. `Extreme16.0_high` (blue) then describes an atmospheric profile with density maximum at an altitude of 16 km (and so on). The results of the simulations, in term of the performance of the observatory in comparison with the profile `Average Winter` – are shown in Figure 3 (energy resolution), Figure 4 (energy bias), Figure 5 (effective area), and Figure 6 (angular resolution). In these figures 'relative' means relative with respect to the profile `Average Winter`. For reconstructions of simulated events we were using the *correct* atmosphere, i.e. the same atmosphere was used for analysis as for the original showers. Note that this is not the usual realistic case when we do not have exact information about the atmosphere (see below). These studies were mainly performed by Federico Di Pierro (INFN Torino) with our participation.

We can see that these atmospheric profiles have relatively small influence on energy resolution of the array, especially in the core energy range (100 GeV ~ 10 TeV) where the error is $\pm 2\%$. In energy bias we do not expect a large effect as we are using the same atmosphere for reconstruction of events as for the data production. The large effects at low energies(< 20 GeV) are due to a selection of the events which suffered large fluctuations. Above 20 GeV this is only $\pm 2\%$ effect. Effective area of telescopes is influenced more by these atmospheric conditions — up to 10% in low and high (≥ 10 TeV) energy range, which leads to uncertainties in the flux. At high energy the atmospheres denser in their higher part (~14 km) produce larger effective areas. The goal angular resolution of the CTA array is around $0.1°$ at 65 GeV, $0.03°$ at 1 TeV and $0.02°$ at 10 TeV, which is satisfied with these profiles. All these analyses where performed with exact profiles, i.e. in case when we have our atmosphere calibrated. But this is not the real case when we have only limited knowledge about the atmosphere. We can also do these analyses without atmospheric calibration, i.e. produce data with extreme profiles and then use the average winter's lookup tables (LUT) to reconstruct data. In this case we expect a large effect on the energy bias but only minor effects on the energy resolution (the spread of the reconstructed energies should not increase). Preliminary result of these simulations are shown in Figure 7 (energy resolution), Figure 8 (energy bias) and Figure 9 (angular resolution). We can see that in energy bias there is an energy dependency which could be dangerous for determining energy spectra. The errors are up to $\pm 4\%$ whereas only 2% systematic shift at all energies is allowed. For energy and angular resolutions these effects are rather small as expected.

In the future the analysis needs to be extended also to the southern site and the effect of different kinds of aerosols (dust, clouds) under different conditions (different altitude and thickness) and of the effect of atmospheric calibration uncertainties on reconstructed energy and flux uncertainties.

4 Conclusion

In the paper we described general status of MC simulations within CTA and in more detail our studies of extreme density profiles for La Palma and their effects on the anticipated performance of the observatory. So far we managed to get results with extreme atmospheric profiles with different density, thickness, refractive index, and optical depth profiles for the northern site. We showed that in case we have precise measurements of the atmosphere we can account for these effects and the performance of the observatory is not greatly influenced. Preliminary results without precise atmospheric calibrations showed that there is a large impact on energy bias and therefore it can greatly influence energy spectra.

We are planning to do similar simulations for the southern site as well as to study effects of different aerosols. We also want to perform these studies at different zenith angle. Results without atmospheric calibrations showed that we need to study the effects of the uncertainties of the currently foreseen CTA atmospheric measurements.

Acknowledgments. This research was supported by Ministry of Education, Youth and Sports, MEYS LE13012, LG14019, LM2015046 and the Czech Science Foundation grant 14-17501S, Czech Republic.

References

[1] Acharya, B. S., et al., *Introducing the CTA concept*, Astroparticle Physics, 43 (2013)

[2] Cherenkov Telescope Array. (2017). CTA Performance - Cherenkov Telescope Array. [online] Available at: https://www.cta-observatory.org/science/cta-performance/ [Accessed 9 Jan. 2017].

[3] Gaug M., *CTA Atmospheric Calibration*,these proceedings.

[4] Heck, D., et al., *Report FZKA 6019*, Forschungszentrum Karlsruhe, http://www-ik.fzk.de/corsika/physics_description/corsika_phys.html, 1998

[5] Bernlöhr, *Simulation of Imaging Atmospheric Cherenkov Telescopes with CORSIKA and* `sim_telarray`, eprint arXiv:0808.2253 (2008).

[6] Maier, G., A short description of an *evnDisplay*-based CTA analysis, https://znwiki3.ifh.de/CTA/Eventdisplay%20Software (Accessed 19 Dec. 2016).

[7] Moralejo, A., Gaug, M., Carmona, E., et al., *MARS, the MAGIC Analysis and Reconstruction Software*, 31st ICRC, Łódź, eprint arXiv:0907.0943 (2009)

[8] Wood, M., Jogler, T., Dumm, J., Funk, S., *Monte Carlo Studies of medium-size telescope designs for the Cherenkov Telescope Array*, Astroparticle Physics 72, eprint arXiv:1506.07476 (2016)

[9] Egi.eu. (2016). EGI | EGI Advanced Computing Services for Research. [online] Available at: https://www.egi.eu/ (Accessed 15 Dec. 2016).

[10] Tsaregorodtsev, A., et al., *DIRAC Distributed Computing Services*, Journal of Physics: Conference Series, 513 (2014)

[11] Arrabito, L., et al., *Prototype of a production system for Cherenkov Telescope Array with DIRAC*, Journal of Physics: Conference Series, 664 (2015)

[12] Arrabito, L., Bregeon, J., *CTA Computing Grid resources and technical aspects of CTA MC simulations*, CCF Face-to-face meeting, Barcelona, Presentation (2016).

[13] Mandát D., *All Sky Camera for CTA Site characterization*, these proceedings

[14] Ebr J., *Aerosol measurements with FRAM telescopes*, these proceedings.

[15] Valore L., *Status of the ARCADE Raman Lidar system*, these proceedings.

[16] Marín J., Cuevas O. and Pozo D., *Density profile characterization and modeling at Paranal and Armazones 2k sites*, these proceedings.

[17] Gaug M., *Site characterization for CTA-N*, these proceedings.

[18] Di Pierro, F., *Atmospheric simulations*, CCF Face-to-face meeting, Barcelona, Presentation (2016).

 DOI: 10.1051/epjconf/201714401014

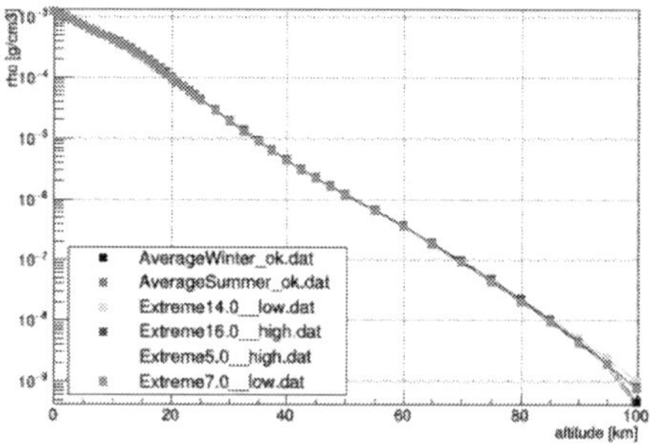

Figure 1. Atmospheric density for different atmospheric profiles (La Palma site). [18]

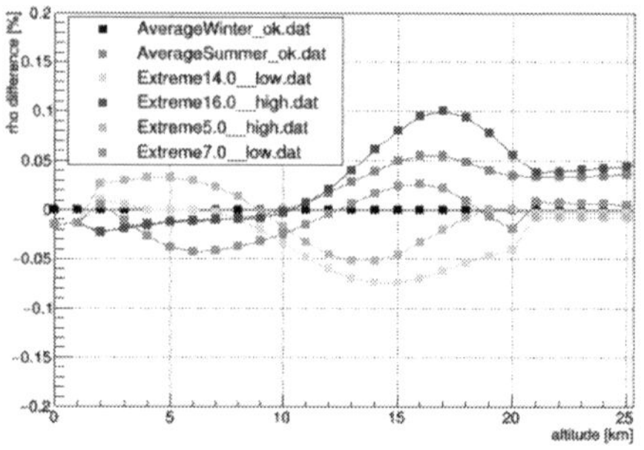

Figure 2. Relative atmospheric density for different atmospheric profiles (La Palma site) with respect to the density profile `Average Winter`. [18]

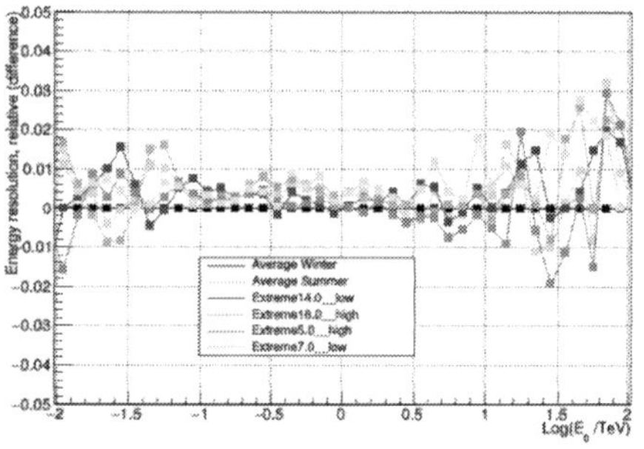

Figure 3. Relative energy resolution for different atmospheric profiles (La Palma site). [18]

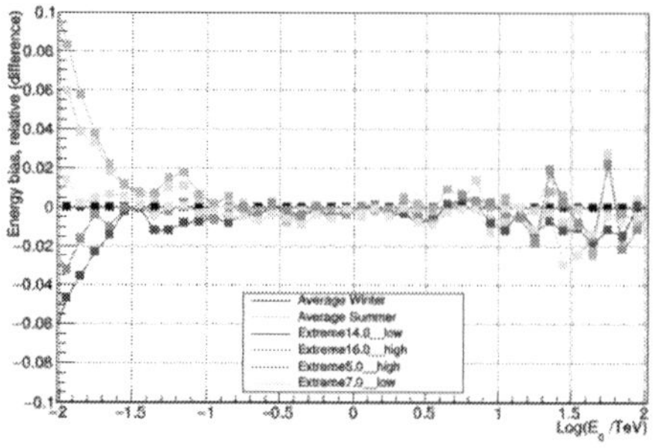

Figure 4. Relative energy bias for different atmospheric profiles (La Palma site). [18]

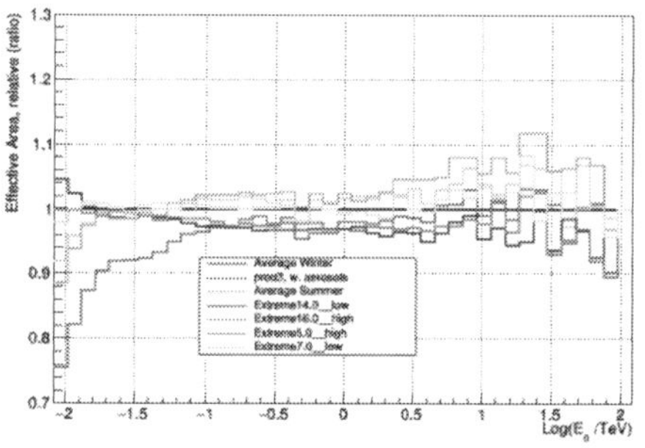

Figure 5. Relative effective area for different atmospheric profiles (La Palma site). [18]

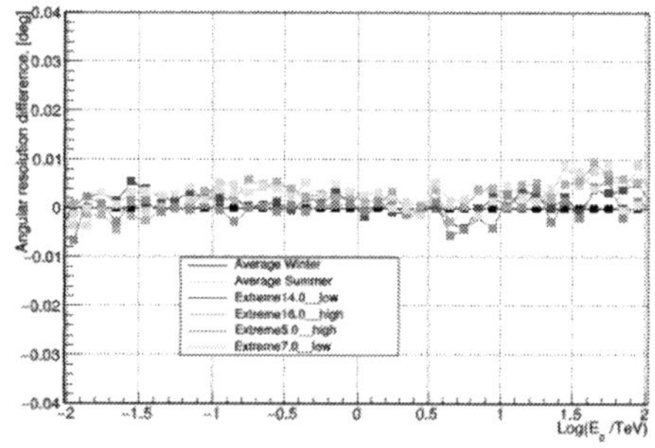

Figure 6. Angular resolution difference for different atmospheric profiles (La Palma site). [18]

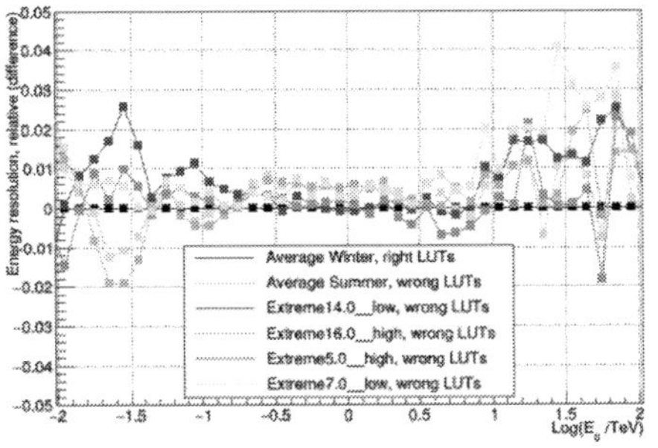

Figure 7. Relative energy resolution for different atmospheric profiles (La Palma site) with wrong LUTs. [18]

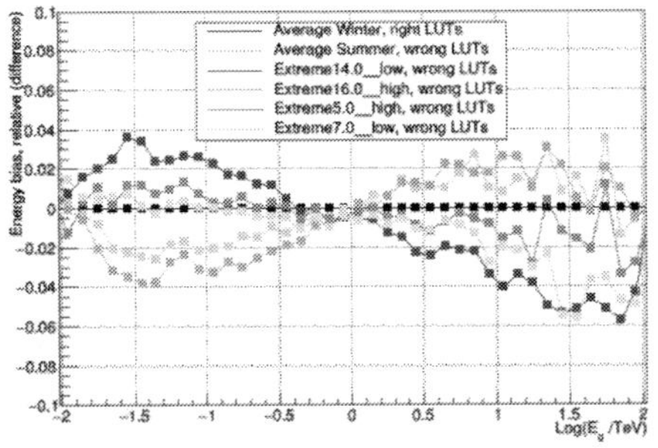

Figure 8. Relative energy bias for different atmospheric profiles (La Palma site) with wrong LUTs. [18]

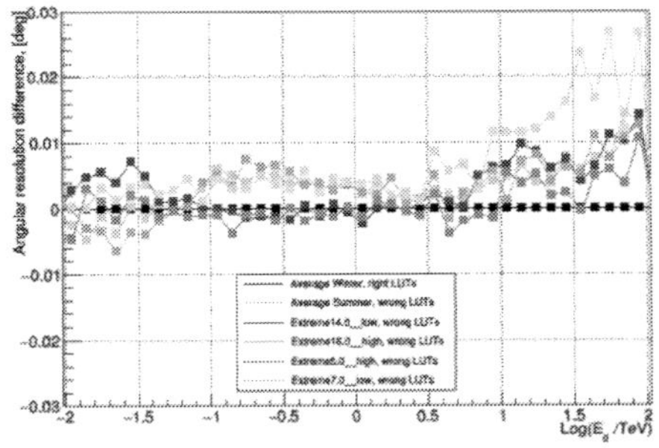

Figure 9. Angular resolution difference for different atmospheric profiles (La Palma site) with wrong LUTs. [18]

AUTHOR INDEX